Fiber Laser and Their Applications

Fiber Laser and Their Applications

Editors

Song Yang
Ling Zhang
Zhiwei Zhu
Li Li

MDPI

Basel • Beijing • Wuhan • Barcelona • Belgrade • Novi Sad • Cluj • Manchester

Editors

Song Yang
Technical University of Denmark
Kgs. Lyngby, Denmark

Ling Zhang
Chinese Academy of Sciences
Beijing, China

Zhiwei Zhu
The Chinese University of Hong Kong
Hong Kong, China

Li Li
Harbin Institute of Technology
Harbin, China

Editorial Office
MDPI
St. Alban-Anlage 66
4052 Basel, Switzerland

This is a reprint of articles from the Special Issue published online in the open access journal *Photonics* (ISSN 2304-6732) (available at: https://www.mdpi.com/journal/photonics/special_issues/ZGM9BWF73N).

For citation purposes, cite each article independently as indicated on the article page online and as indicated below:

Lastname, A.A.; Lastname, B.B. Article Title. *Journal Name* **Year**, *Volume Number*, Page Range.

ISBN 978-3-0365-8713-4 (Hbk)
ISBN 978-3-0365-8712-7 (PDF)
doi.org/10.3390/books978-3-0365-8712-7

Contents

Preface

Fiber lasers are key components that have attracted wide attention owing to their advantages such as compact structure, high peak power, high efficiency, and excellent beam quality. To date, considerable efforts have been undertaken to improve the emission performance of optical fiber lasers, including their pulse duration, peak power, and stability. Furthermore, high-peak-power fiber lasers have been adopted as an edge tool in both research and industry fields, such as nonlinear optics, optical imaging, laser processing, and 3D printing. The aim of this reprint is to provide a comprehensive exploration of fiber lasers and their diverse applications.

Song Yang, Ling Zhang, Zhiwei Zhu, and Li Li
Editors

photonics

Communication

Development of a Mode-Locked Fiber Laser Utilizing a Niobium Diselenide Saturable Absorber

Weiqin Guo [1,2], Ling Zhang [2], Xiaosheng Xiao [3,*], Xingxing Li [1], Zhigang Yin [2,4], Hui Ning [1,2], Xin Zhang [1] and Xingwang Zhang [2,4]

[1] Joint Lab of Digital Optical Chip, Wuyi University, Jiangmen 529020, China; weiqinguo2023@163.com (W.G.); wyuxxli@163.com (X.L.); nh13073763152@163.com (H.N.); zhangxin800800@163.com (X.Z.)
[2] Key Lab of Semiconductor Materials Science, Institute of Semiconductors, Chinese Academy of Sciences, Beijing 100083, China; zhangling@semi.ac.cn (L.Z.); yzhg@semi.ac.cn (Z.Y.); xwzhang@semi.ac.cn (X.Z.)
[3] School of Electronic Engineering, Beijing University of Posts and Telecommunications, Beijing 100876, China
[4] Center of Materials Science and Optoelectronics Engineering, University of Chinese Academy of Sciences, Beijing 100049, China
* Correspondence: xsxiao@bupt.edu.cn

Abstract: The saturable absorber of niobium diselenide ($NbSe_2$), with a wide working bandwidth and excellent nonlinear optical response, was prepared using liquid-phase exfoliation. Its saturation intensity and modulation depth were 5.35 MW/cm^2 and 6.3%, respectively. Stable mode-locking with a center wavelength of 1558.7 nm of an erbium-doped fiber laser based on a $NbSe_2$ saturable absorber was successfully achieved. The maximum average output power of the mode-locked laser was 6.93 mW, with a pulse width of 1.3 ps and a repetition rate of 25.31 MHz at a pump power of 550 mW. The results show that $NbSe_2$ is a promising photoelectric modulation material owing to its excellent nonlinear optical properties.

Keywords: niobium diselenide; saturable absorber; nonlinear absorption characteristics; fiber laser; mode-locked laser

Citation: Guo, W.; Zhang, L.; Xiao, X.; Li, X.; Yin, Z.; Ning, H.; Zhang, X.; Zhang, X. Development of a Mode-Locked Fiber Laser Utilizing a Niobium Diselenide Saturable Absorber. *Photonics* **2023**, *10*, 610. https://doi.org/10.3390/photonics10060610

Received: 27 April 2023
Revised: 19 May 2023
Accepted: 22 May 2023
Published: 25 May 2023

1. Introduction

Ultrashort pulse lasers have been widely applied in medical diagnosis, optical detection, material processing, precision machining and fiber communication due to their advantages of a narrow pulse width and high peak power [1–3]. Passive mode-locking technology is the main method used to achieve ultrashort pulse output. Compared with other lasers, fiber laser has the advantages of a low cost, compact structure, high efficiency, easy integration and good stability. Therefore, passive-mode-locked fiber lasers have become a research hotspot [4–6].

The passive mode-locked laser cannot be realized without the help of a saturable absorber (SA). Although the traditional semiconductor saturable absorption mirror (SESAM) has a mature preparation technology and high stability, it is expensive and has a narrow working band [7]. Therefore, it is necessary to explore new SAs. The detailed study of graphene material has aroused research interest in SAs based on two-dimensional (2D) materials. Researchers have found that many 2D materials exhibit special linear and nonlinear (e.g., saturable absorption) properties in terms of optics, ranging from visible to mid-infrared light. This research shows that 2D materials have many potential applications in optical devices [8–11]. Exploring 2D materials with excellent optical properties is a key factor in developing high-performance pulsed lasers [12].

To date, 2D materials with saturable absorption have been discovered, including graphene [13,14], black phosphorus (BP) [15,16], topological insulators (TIs) [17,18], MXenes [19,20] and transition metal dihalides (TMDs) [21–24]. Among them, TMDs have garnered considerable attention due to their fast saturation recovery time and large modulation depth [25–28]. TMDs have a chemical formula of MX_2, where M represents transition metal elements (W, Mo, Nb, Ta) and X represents chalcogen elements (S, Se, Te) [29–32]. In recent years, $NbSe_2$ has become a hot research topic in TMDs due to its unique properties, such as an excellent photoelectric response, high carrier mobility and superconductivity [33–35], making it a popular candidate material for optoelectronic device applications. In 2018, Shi et al. deposited $NbSe_2$ quantum dots onto a D-type fiber to achieve mode-locked lasers that operate at 1 and 1.5 μm [36]. In 2020, Chen et al. prepared a $NbSe_2$ saturable absorber using the mechanical stripping method and realized a mode-locked pulse output with a central wavelength of 1036 nm and a pulse width of 174 ps [37]. In the same year, Yang obtained a mode-locked laser with a pulse width of 697 fs and a central wavelength of 1556.3 nm using an optical deposition method [38].

In this paper, we prepared an $NbSe_2$ SA using liquid-phase exfoliation (LPE). The nonlinear test results show that the saturation intensity, modulation depth and nonsaturable absorption were 5.35 MW/cm^2, 6.3% and 67.18%, respectively. We inserted the SA into an erbium-doped fiber laser, successfully achieving a stable mode-locked pulse laser output at pump powers ranging from 90 to 550 mW. At a pump power of 550 mW, the pulse width, repetition rate and average output power were 1.3 ps, 25.31 MHz and 6.93 mW, respectively. The center wavelength of the mode-locked laser was 1558.7 nm. The root mean square (RMS) instability of the mode-locked laser at the pump power of 550 mW was less than 1%. We achieved a higher output power and more stable laser output than other reported results of mode-locked lasers based on $NbSe_2$ SA. The experimental results show that $NbSe_2$ has excellent nonlinear optical properties and future applications in ultrafast photonics devices.

2. Preparation and Characterization

In our experiment, $NbSe_2$ nanosheets were prepared using the LPE method due to its low cost, simple operation and easy realization of large-scale preparation. Firstly, we added 50 mg of sodium dodecyl sulfate (SDS) and 300 mg of polyvinyl alcohol (PVA) to 20 mL of deionized water (DI), heated it to 90 °C, and kept it for half an hour to fully dissolve the SDS and PVA. Secondly, we added 100 mg of $NbSe_2$ powder to the above solution and placed it in an ultrasonic cleaning machine for 10 h. Finally, the mixed solution was centrifuged at a speed of 5000 rpm for 1 h, and $\frac{1}{2}$ of the supernatant was taken to make the $NbSe_2$ nanosheet solution used in our experiment. The nanosheet solution was dropped onto the fiber jumper and the fiber was dried in the drying cabinet for 24 h. As shown in Figure 1, we deposited $NbSe_2$-PVA on the end face of the fiber jumper.

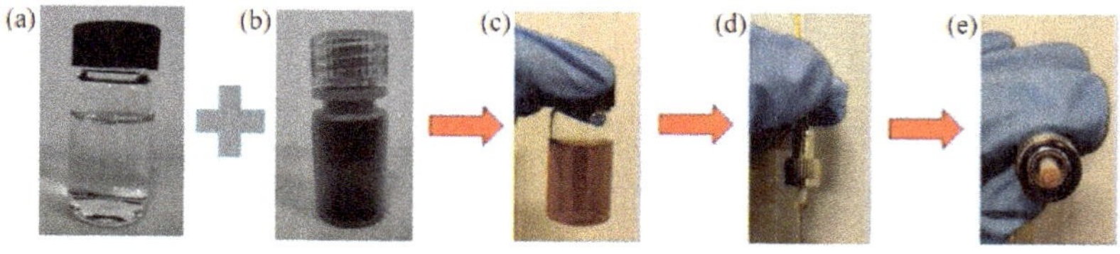

Figure 1. Preparation process of $NbSe_2$-PVA SA. (**a**) SDS and PVA mixture. (**b**) $NbSe_2$ powder. (**c**) $NbSe_2$-PVA dispersion. (**d**) Dispersion was dropped onto fiber end face. (**e**) $NbSe_2$-PVA was deposited on fiber end face.

The nanosheet solution was spin-coated onto the silicon substrate for characterization. Raman spectroscopy was used to characterize $NbSe_2$ nanosheets, and the corresponding results are presented in Figure 2. Figure 2a illustrates that two peaks were observed at 227.7 and 238.5 cm^{-1}, which correspond to the A_{1g} and $E^1{}_{2g}$ vibrational modes of $NbSe_2$, respectively. These findings are similar to those reported previously [38]. The scanning electron microscopy (SEM) image of the $NbSe_2$ nanosheets is shown in Figure 2b. $NbSe_2$ nanoparticles were uniformly distributed on the substrate. Furthermore, atomic force microscopy (AFM) was employed to characterize the material. As illustrated in Figure 2c, the thickness of $NbSe_2$ nanosheets was about 50 nm according to the marked line profiles.

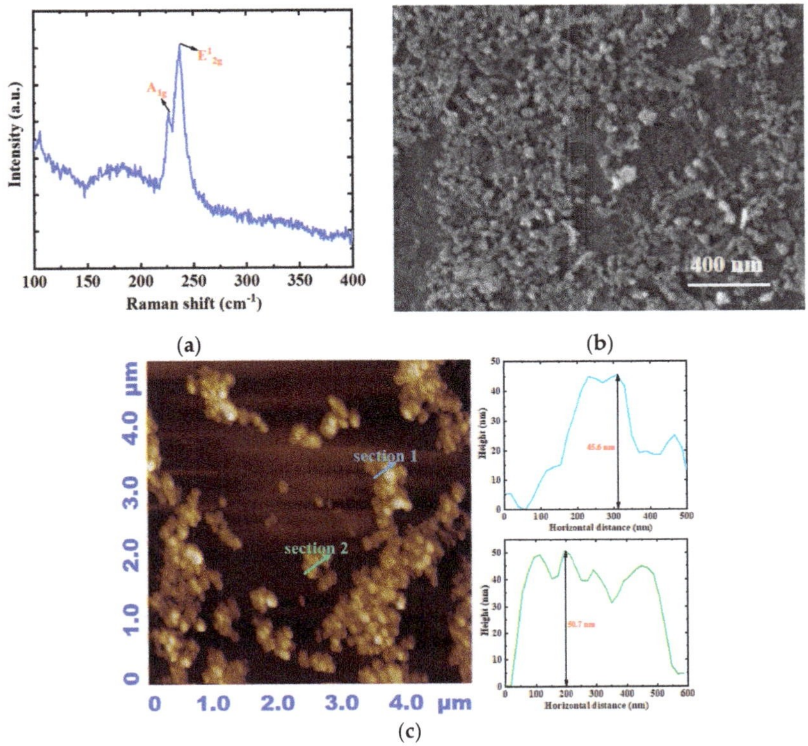

Figure 2. (**a**) Raman spectrum of $NbSe_2$ nanosheets. (**b**) SEM image of $NbSe_2$ nanosheets. (**c**) AFM image of $NbSe_2$ nanosheets.

A technique using a balanced twin-detector measurement was used to measure the saturable absorption characteristics of $NbSe_2$ nanosheets. The experimental setup is illustrated in Figure 3a. The pump laser, with a tunable wavelength in the near-infrared range, has a pulse duration of 300 fs and a repetition rate of 75 kHz. The experimental results were fitted using a non-linear transmission function. The relationship between light transmittance T and incident light intensity I can be described as follows [39]

$$T = 1 - \frac{\Delta T}{1 + I/I_{sat}} - T_{\rm ns} \qquad (1)$$

where ΔT is the modulation depth, $I_{\rm sat}$ is the saturation intensity, and $T_{\rm ns}$ is the nonsaturable absorption. The dependence of transmittance on the pump fluence is shown in Figure 3b. According to the fitting results, the modulation depth, saturation intensity and nonsaturable loss of $NbSe_2$ SA are determined to be 6.3%, 5.35 MW/cm^2 and 67.18%, respectively.

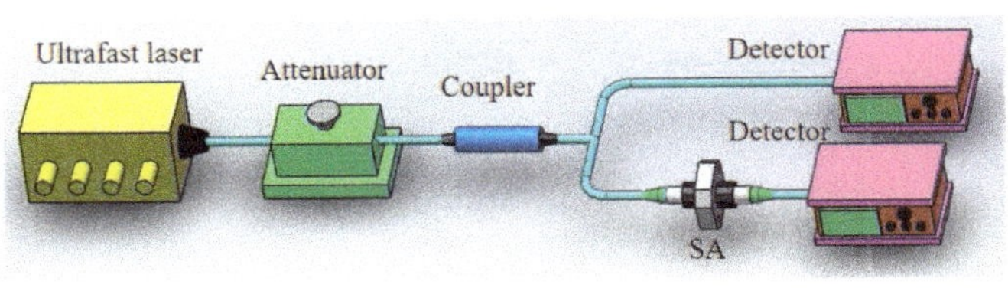

(a)

Figure 3. *Cont.*

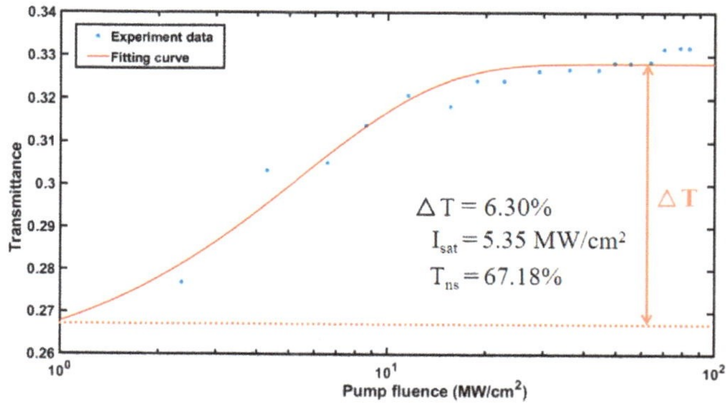

(**b**)

Figure 3. (**a**) Balanced twin-detector measurement system. (**b**) Nonlinear transmittance diagram of $NbSe_2$ nanosheets.

3. Results and Discussion

The experimental setup of mode-locked fiber laser based on $NbSe_2$-PVA SA is depicted in Figure 4. A ring cavity structure was adopted, including a pump source (a 976 nm laser diode LD), 976/1550 nm wavelength division multiplexer (WDM), erbium-doped fiber (EDF), polarization controller (PC), polarization-independent isolator (PI-ISO), $NbSe_2$-PVA SA, single-mode fiber and 90/10 output coupler (OC). The pump light was coupled to the laser cavity through 976/1550 nm WDM. A 0.7 m single-mode erbium-doped fiber (Er80-8/125) was employed as the laser gain medium. PC was applied to adjust the polarization state of the laser. We used PI-ISO to ensure the unidirectional propagation of light in the laser cavity. The 90/10 OC was applied to the output laser. The total length of the ring laser cavity is about 8 m.

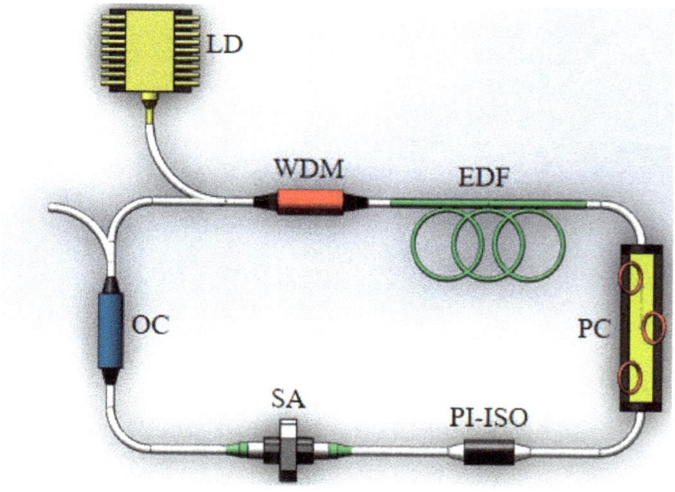

Figure 4. Erbium-doped mode-locked fiber laser based on $NbSe_2$-PVA SA.

During the mode-locked experiment, the PC was adjusted to change the polarization state in the cavity. The laser pulse was observed using 5 GHz photodiode (Thorlabs, DET08CFC/M) and 500 MHz real-time oscilloscope (KEYSIGHT DSOX3052T). At a pump power of 90 mW, the laser changed from continuous light to a mode-locked pulse, which was observed using the real-time oscilloscope. A stable mode-locked laser was obtained as the pump power was further increased, from 90 to 550 mW.

At a pump power of 550 mW, the pulse sequence diagram of the mode-locked laser observed by the oscilloscope is illustrated in Figure 5a. The pulse interval is about 39.5 ns, corresponding to a repetition rate of 25.31 MHz, which is consistent with the result estimated from the cavity length. Then, the mode-locked laser spectrum was measured by a spectral analyzer (Agilent 86142B) with a resolution of 0.06 nm, and the result is demonstrated in Figure 5b. The center wavelength of the mode-locked pulse is 1558.7 nm, and the 3 dB bandwidth is 4.69 nm. The mode-locked laser was measured with an autocorrelator (A.P.E. pulseCheck), and the data are shown in Figure 5c by the $sech^2$ fitting. The full width at half maximum (FWHM) of the autocorrelation trace is 2 ps, which means that the pulse duration is 1.3 ps, assuming a hyperbolic secant shape. The calculated time-bandwidth product (TBP) is 0.753, whereas the theoretical limit value of TBP is 0.315. Our experimental results indicate that the mode-locked pulse has some chirp. Figure 5d shows the radio frequency (RF) spectrum measured by an RF signal analyzer (Agilent N9020 A). The repetition rate of the mode-locking laser is 25.31 MHz, and the signal-to-noise ratio (SNR) reaches 80 dB, indicating a highly stable mode-locking operation. Figure 5e demonstrates the relationship between the output power and pump power. At a pump power of 550 mW, the output power is 6.93 mW and the corresponding optical conversion efficiency is calculated to be 1.26%. The calculated single pulse energy and peak power are about 0.27 nJ and 207.7 W, respectively. To test the stability of the laser, the output power curve of the mode-locked laser for 30 min at a pump power of 550 mW was measured, as shown in Figure 5f. The calculated RMS instability of the output power of the mode-locked laser is less than 1%.

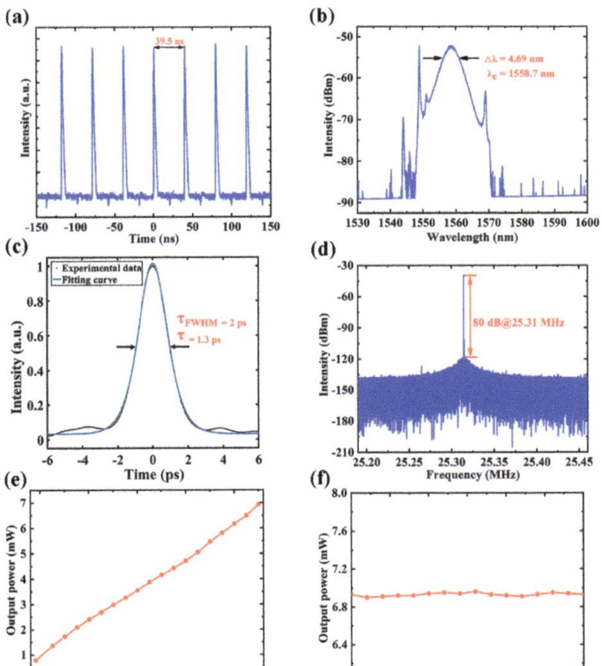

Figure 5. Output characteristics of the laser at 550 mW pump power. (**a**) Pulse train. (**b**) Spectrogram. (**c**) Autocorrelation trace. (**d**) RF spectrum. (**e**) Relationship diagram between the output power and pump power. (**f**) Stability of output power of mode-locked laser at pump power of 550 mW.

Removing the $NbSe_2$-PVA SA from the cavity and observing that no mode-locked pulsed laser is generated, even after adjusting the polarization controller and changing the pump power, confirms that the mode-locked operation is indeed caused by the $NbSe_2$-PVA SA.

4. Conclusions

In summary, we prepared an $NbSe_2$-PVA saturable absorber using LPE and inserted it into a homemade erbium-doped fiber laser. The laser produced stable mode-locked pulses with an output power of 6.93 mW, a center wavelength of 1558.7 nm, a repetition rate of 25.31 MHz, and a pulse width of 1.3 ps at a pump power of 550 mW. The output power of the mode-locked laser was measured for 30 min at the maximum pump power. The RMS instability of the output power was less than 1%. The results suggest that $NbSe_2$ has excellent nonlinear optical properties and promising applications in the field of ultrafast lasers.

Author Contributions: Conceptualization, X.X.; methodology, X.L.; investigation, W.G. and H.N.; resources, L.Z.; data curation, W.G. and L.Z.; writing—original draft preparation, W.G and X.Z. (Xin Zhang); writing—review and editing, W.G., L.Z. and X.X.; supervision, X.Z. (Xingwang Zhang) and Z.Y.; project administration, L.Z. and X.L. All authors have read and agreed to the published version of the manuscript.

Funding: This work is financially supported by Beijing Municipal Science & Technology Commission, Administrative Commission of Zhongguancun Science Park (No. Z211100004821006).

Institutional Review Board Statement: Not applicable.

Informed Consent Statement: Not applicable.

Data Availability Statement: The data of this study are available from the corresponding author upon reasonable request.

Acknowledgments: The authors would like to thank Chenxin Gao of Department of Precision Instruments, Tsinghua University for help on the characterization of the SA.

Conflicts of Interest: The authors declare no conflict of interest.

References

1. Ahmad, H.; Muhammad, F.D.; Pua, C.H.; Thambiratnam, K. Dual-Wavelength Fiber Lasers for the Optical Generation of Microwave and Terahertz Radiation. *IEEE J. Sel. Top. Quantum Electron.* **2014**, *20*, 166–173. [CrossRef]
2. Hu, P.; Mao, J.; Nie, H.; Wang, R.; Zhang, B.; Li, T.; He, J.; Yang, K. Highly Stable Passively Q-Switched Erbium-Doped All-Fiber Laser Based on Niobium Diselenide Saturable Absorber. *Molecules* **2021**, *26*, 4303. [CrossRef]
3. Zheng, J.-C.; Yang, S.; Zhu, Z.-W.; Lau, K.-Y.; Li, L. 72-fs Er-doped Mamyshev Oscillator. *J. Light. Technol.* **2021**, *40*, 2123–2127. [CrossRef]
4. Zhao, L.; Tang, D.; Wu, X.; Zhang, H.; Lu, C.; Tam, H.Y. Observation of dip-type sidebands in a soliton fiber laser. *Opt. Commun.* **2010**, *283*, 340–343. [CrossRef]
5. Yang, S.; Zhang, Q.-Y.; Zhu, Z.-W.; Qi, Y.-Y.; Yin, P.; Ge, Y.-Q.; Li, L.; Jin, L.; Zhang, L.; Zhang, H. Recent advances and challenges on dark solitons in fiber lasers. *Opt. Laser Technol.* **2022**, *152*, 108116. [CrossRef]
6. Yang, S.; Li, F.; Gong, M.-M.; Zhang, L.; Zhu, Z.-W.; Shen, H.-B.; Chen, S.-C. Generation of Q-switched and mode-locked pulses based on PbS/CdS saturable absorbers in an Er-doped fiber laser. *J. Mater. Chem. C* **2022**, *10*, 5956–5961. [CrossRef]
7. Gomes, L.; Orsila, L.; Jouhti, T.; Okhotnikov, O. Picosecond SESAM-Based Ytterbium Mode-Locked Fiber Lasers. *IEEE J. Sel. Top. Quantum Electron.* **2004**, *10*, 129–136. [CrossRef]
8. Novoselov, K.S.; Geim, A.K.; Morozov, S.V.; Jiang, D.; Katsnelson, M.I.; Grigorieva, I.V.; Dubonos, S.V.; Firsov, A.A. Two-dimensional gas of massless Dirac fermions in graphene. *Nature* **2005**, *438*, 197–200. [CrossRef]
9. Zhang, Y.B.; Tan, Y.W.; Stormer, H.L.; Kim, P.H. Experimental observation of the quantum Hall effect and Berry's phase in graphene. *Nature* **2005**, *438*, 201–204. [CrossRef]
10. Mak, K.F.; Lee, C.G.; Hone, J.; Shan, J.; Heinz, T.J. Atomically thin MoS_2: A new direct-gap semiconductor. *Phys. Rev. Lett.* **2010**, *105*, 136805. [CrossRef]
11. Qi, Y.; Yang, S.; Wang, J.; Li, L.; Bai, Z.; Wang, Y.; Lv, Z. Recent advance of emerging low-dimensional materials for vector soliton generation in fiber lasers. *Mater. Today Phys.* **2022**, *23*, 100622. [CrossRef]
12. Lu, S.; Wen, S.; Du, L.; Kang, Z.; Li, J.; Huang, B.; Jiang, G.; Miao, L.; Qin, G.; Zhao, C. Stable Dissipative Soliton Generation From Yb-Doped Fiber Laser Modulated via Evanescent Field Interaction With Gold Nanorods. *IEEE Photon. J.* **2018**, *10*, 1–8. [CrossRef]

13. Zhang, H.; Tang, D.; Knize, R.J.; Zhao, L.; Bao, Q.; Loh, K.P. Graphene mode locked, wavelength-tunable, dissipative soliton fiber laser. *Appl. Phys. Lett.* **2010**, *96*, 111112. [CrossRef]
14. Bonaccorso, F.; Sun, Z.; Hasan, T.; Ferrari, A.C. Graphene photonics and optoelectronics. *Nat. Photon* **2010**, *4*, 611–622. [CrossRef]
15. Xia, F.; Wang, H.; Jia, Y. Rediscovering black phosphorus as an anisotropic layered material for optoelectronics and electronics. *Nat. Commun.* **2014**, *5*, 4458. [CrossRef]
16. Sotor, J.; Sobon, G.; Kowalczyk, M.; Macherzynski, W.; Paletko, P.; Abramski, K.M. Ultrafast thulium-doped fiber laser mode locked with black phosphorus. *Opt. Lett.* **2015**, *40*, 3885–3888. [CrossRef]
17. Zhao, C.; Zou, Y.; Chen, Y.; Wang, Z.; Lu, S.; Zhang, H.; Wen, S.; Tang, D. Wavelength-tunable picosecond soliton fiber laser with Topological Insulator: Bi_2Se_3 as a mode locker. *Opt. Express* **2012**, *20*, 27888–27895. [CrossRef]
18. Liu, H.; Zheng, X.-W.; Liu, M.; Zhao, N.; Luo, A.-P.; Luo, Z.-C.; Xu, W.-C.; Zhang, H.; Zhao, C.-J.; Wen, S.-C. Femtosecond pulse generation from a topological insulator mode-locked fiber laser. *Opt. Express* **2014**, *22*, 6868–6873. [CrossRef]
19. Dong, Y.C.; Chertopalov, S.; Maleski, K.; Anasori, B.; Hu, L.; Bhattacharya, S.; Rao, A.M.; Gogotsi, Y.; Mochalin, V.N.; Podila, R. Saturable absorption in 2D Ti_3C_2 MXene thin films for passive photonic diodes. *Adv. Mater.* **2018**, *30*, 1705714. [CrossRef]
20. Wu, Q.; Jin, X.; Chen, S.; Jiang, X.; Hu, Y.; Jiang, Q.; Wu, L.; Li, J.; Zheng, Z.; Zhang, M.; et al. MXene-based saturable absorber for femtosecond mode-locked fiber lasers. *Opt. Express* **2019**, *27*, 10159–10170. [CrossRef]
21. Du, J.; Wang, Q.; Jiang, G.; Xu, C.; Zhao, C.; Xiang, Y.; Chen, Y.; Wen, S.; Zhang, H. Ytterbium-doped fiber laser passively mode locked by few-layer Molybdenum Disulfide (MoS_2) saturable absorber functioned with evanescent field interaction. *Sci. Rep.* **2014**, *4*, srep06346. [CrossRef] [PubMed]
22. Zhang, H.; Lu, S.B.; Zheng, J.; Du, J.; Wen, S.C.; Tang, D.Y.; Loh, K.P. Molybdenum disulfide (MoS_2) as a broadband saturable absorber for ultra-fast photonics. *Opt. Express* **2014**, *22*, 7249–7260. [CrossRef] [PubMed]
23. Xia, H.; Li, H.; Lan, C.; Li, C.; Zhang, X.; Zhang, S.; Liu, Y. Ultrafast erbium-doped fiber laser mode-locked by a CVD-grown molybdenum disulfide (MoS_2) saturable absorber. *Opt. Express* **2014**, *22*, 17341–17348. [CrossRef] [PubMed]
24. Woodward, R.I.; Howe, R.C.T.; Runcorn, T.H.; Hu, G.; Torrisi, F.; Kelleher, E.J.R.; Hasan, T. Wideband saturable absorption in few-layer molybdenum diselenide ($MoSe_2$) for Q-switching Yb-, Er-and Tm-doped fiber lasers. *Opt. Express* **2015**, *23*, 20051. [CrossRef]
25. Feldman, Y.; Wasserman, E.; Srolovitz, D.J.; Tenne, R. High-rate, gas-phase growth of MoS_2 nested inorganic fullerenes and nanotubes. *Science* **1995**, *267*, 222. [CrossRef]
26. Mao, D.; Du, B.; Yang, D.; Zhang, S.; Wang, Y.; Zhang, W.; She, X.; Cheng, H.; Zeng, H.; Zhao, J. Nonlinear Saturable Absorption of Liquid-Exfoliated Molybdenum/Tungsten Ditelluride Nanosheets. *Small* **2016**, *12*, 1489–1497. [CrossRef]
27. Woodward, R.I.; Kelleher, E.J.R. 2D Saturable Absorbers for Fibre Lasers. *Appl. Sci.* **2015**, *5*, 1440–1456. [CrossRef]
28. Yang, H.; Wang, Y.; Tiu, Z.C.; Tan, S.J.; Yuan, L.; Zhang, H. All-Optical Modulation Technology Based on 2D Layered Materials. *Micromachines* **2022**, *13*, 92. [CrossRef]
29. Luo, Z.; Wu, D.; Xu, B.; Xu, H.; Cai, Z.; Peng, J.; Weng, J.; Xu, S.; Zhu, C.; Wang, F.; et al. Two-dimensional material-based saturable absorbers: Towards compact visible-wavelength all-fiber pulsed lasers. *Nanoscale* **2016**, *8*, 1066–1072. [CrossRef]
30. Navarro-Moratalla, E.; Island, J.O.; Mañas-Valero, S.; Pinilla-Cienfuegos, E.; Castellanos-Gomez, A.; Quereda, J.; Rubio-Bollinger, G.; Chirolli, L.; Silva-Guillén, J.A.; Agraït, N.; et al. Enhanced superconductivity in atomically thin TaS2. *Nat. Commun.* **2016**, *7*, 11043. [CrossRef]
31. Zhao, Q.Y.; Guo, Y.H.; Zhou, Y.X.; Xu, X.; Ren, Z.; Bai, J.; Xu, X. Flexible and anisotropic properties of monolayer MX2 (M = Tc and Re; X = S, Se). *J. Phys. Chem. C* **2017**, *121*, 23744. [CrossRef]
32. Yin, J.; Zhu, F.; Lai, J.; Chen, H.; Zhang, M.; Zhang, J.; Wang, J.; He, T.; Zhang, B.; Yuan, J.; et al. Hafnium Sulfide Nanosheets for Ultrafast Photonic Device. *Adv. Opt. Mater.* **2019**, *7*, 1801303. [CrossRef]
33. Bachmann, R.; Kirsch, H.; Geballe, T. Optical properties and superconductivity of NbSe2. *Solid State Commun.* **1971**, *9*, 57–60. [CrossRef]
34. Huang, Y.H.; Chen, R.S.; Zhang, J.R. Electronic transport in $NbSe_2$two-dimensional nanostructures: Semiconducting characteristics and photoconductivity. *Nanoscale* **2015**, *7*, 18964–18970. [CrossRef] [PubMed]
35. Sohn, E.; Xi, X.; He, W.-Y.; Jiang, S.; Wang, Z.; Kang, K.; Park, J.-H.; Berger, H.; Forró, L.; Law, K.T.; et al. An unusual continuous paramagnetic-limited superconducting phase transition in 2D NbSe 2. *Nat. Mater.* **2018**, *17*, 504–508. [CrossRef]
36. Shi, Y.; Long, H.; Liu, S.; Tsang, Y.H.; Wen, Q. Ultrasmall 2D $NbSe_2$ based quantum dots used for low threshold ultrafast lasers. *J. Mater. Chem. C* **2018**, *6*, 12638–12642. [CrossRef]
37. Chen, L.; Du, L.; Li, J.; Yang, L.; Yi, Q.; Zhao, C. Dissipative Soliton Generation From Yb-Doped Fiber Laser Modulated by Mechanically Exfoliated NbSe2. *Front. Phys.* **2020**, *8*, 320. [CrossRef]
38. Yang, H. Niobium diselenide nanosheets for a vector soliton fiber laser. *J. Mater. Chem. C* **2020**, *8*, 14954–14958. [CrossRef]
39. Jin, L.; Zhang, Q.; Zhang, B.; Gao, Z.; Yang, S.; Li, L. Numerical analysis of hybrid mode-locking stability in a Ho-doped fiber laser. *Opt. Express* **2023**, *31*, 1141. [CrossRef]

photonics

Article

Orthogonally Polarized Dual-Wavelength Gain-Switched Ho:LuLiF$_4$ Pulse Laser

Guanqu Hu [1], Jinhui Cui [1], Fengjun Tian [1], Zhengxin Gao [1], Shixiong Yan [1], Sichen Liu [1], Xinlu Zhang [2,*] and Li Li [1,3,*]

1 Key Laboratory of In-Fiber Integrated Optics of Ministry of Education, Harbin Engineering University, Harbin 150001, China
2 School of Physical Science and Technology, Tiangong University, Tianjin 300387, China
3 School of Physics, Harbin Institute of Technology, Harbin 150001, China
* Correspondence: zhangxinlu@tiangong.edu.cn (X.Z.); lylee_heu@hrbeu.edu.cn (L.L.)

Abstract: A compact, orthogonally polarized, gain-switched a-cut Ho:LuLiF$_4$ laser with intra-cavity pumping by a self-Q-switched Tm:YAP laser is demonstrated here for the first time. The π-polarization laser at 2052 nm and σ-polarization laser at 2066 nm were experimentally observed with the maximum output power values of 299 mW and 126 mW, respectively, and the two polarization directions were always kept mutually orthogonal as the pump power increased. The ratio of the output power between the two orthogonal polarization lasers was nearly 1:1 at a pump power of 18.4 W. The minimum pulse width of the Ho:LLF laser was 326 ns, the maximum repetition rate was 24 kHz, and the maximum average energy was 28 μJ.

Keywords: orthogonally polarized laser; Ho:LuLiF$_4$; dual-wavelength

Citation: Hu, G.; Cui, J.; Tian, F.; Gao, Z.; Yan, S.; Liu, S.; Zhang, X.; Li, L. Orthogonally Polarized Dual-Wavelength Gain-Switched Ho:LuLiF$_4$ Pulse Laser. *Photonics* **2023**, *10*, 62. https://doi.org/10.3390/photonics10010062

Received: 5 December 2022
Revised: 31 December 2022
Accepted: 4 January 2023
Published: 6 January 2023

1. Introduction

Ultrafast lasers operating in the 2 μm eye-safe spectrum waveband have attracted much attention for a variety of applications in differential absorption lidars, coherent Doppler wind lidars, remote sensing, laser medicine, and nonlinear frequency conversion in the mid-infrared band [1–6]. Both Tm^{3+} and Ho^{3+} ion-doped lasers are able to generate lasers in the 2 μm band. Compared with Tm-doped lasers, Ho-doped lasers can achieve higher optical-to-optical conversion efficiencies, theoretically close to 95%. The optical-to-optical conversion efficiency of single-doped Ho lasers is also significantly higher than that of the widely used Tm and Ho co-doped lasers (the theoretical limit is below 45%) [7]. Single Ho-doped laser crystals have the advantage of high-energy storage, which provides a reliable guarantee of high-energy laser output. Compared with single Tm-doped and Tm–Ho co-doped laser crystals, single Ho-doped laser crystals possess a larger excited emission cross-section (the emission cross-section of Ho ions is about five times larger than that of Tm ions), a long fluorescence lifetime, and a long upper energy level lifetime (~10–16 ms), which facilitates energy storage for Q-switched laser output and is less likely to cause damage to materials [8–11]. Among the Ho-doped crystals that have received widespread attention such as aluminate crystals (YAG/YAP), silicate crystals (Re_2SiO_5), tungstate crystals ($KY(WO_4)_2$), and fluoride crystals ($LiYF_4$), fluoride crystals usually have smaller phonon energies and larger energy level lifetimes. Ho:LuLiF$_4$ (Ho:LLF) crystals, as members of the fluoride crystal family, can be used in excellent single Ho-doped lasers. In Ho:LLF, the larger ionic radius of the matrix Lu^{3+} ions can generate a stronger crystal field and induce the doped Ho ions to form a larger amount of energy level splitting, which results in Ho:LLF having a wider spectral range and a lower laser threshold [7,12]. Ho:LLF crystals belong to the tetragonal crystal system and possess two polarization directions, namely the electric field E//a axis of the σ-polarization direction and the π-polarization direction of the E⊥a axis. Both directions can produce linearly polarized laser output.

LLF crystals have higher natural birefringence characteristics and lower upconversion efficiency in lasers compared with YAP and YAG crystals [13]. Additionally, Ho:LLF crystals have good thermo-optical and thermomechanical properties. For Ho:LLF crystals, the thermal conductivity (a-axis) is 4.3 $Wm^{-1}K^{-1}$, the linear expansion coefficient (c-axis) is 11×10^{-6} K^{-1}, and the thermal coefficient of refractive index dn/dT (c-axis) is -3.6×10^{-6} K^{-1} [7]. Ho:LLF crystals have many features such as a low upconversion loss, negative thermo-optical coefficient, high resistance to optical damage, and no thermally induced birefringence. These excellent features make Ho:LLF crystals attractive for applications in 2 μm lasers and optical amplifiers [14]. Recently, Ho:LLF lasers operating in the continuous wave and Q-switched modes were studied [15,16]. However, the co-existence of dual wavelengths for orthogonal polarizations in a gain-switched Ho:LLF pulse laser has not been investigated.

An orthogonally polarized laser oscillates two adjacent longitudinal modes with a mutually perpendicular polarization state, with the wavelength (or frequency) difference between the two orthogonally polarized longitudinal modes resulting in a dual-wavelength laser. An orthogonally polarized dual-wavelength laser has a remarkable physical characteristic: the laser wavelength (or frequency) interval can be continuously tuned from tens of MHz to one longitudinal mode interval (several GHz) or even more. Dual-wavelength lasers have been extensively and intensively used in lidars, terahertz research, nonlinear frequency conversion and the generation of multiple types of solitons in fiber lasers, precision laser spectroscopy, and medical applications [6,17,18]. Orthogonal polarization lasers have been a hot item of international interest. In 2010, X. P. Yan et al. developed a 26.2 W orthogonally linearly polarized $Nd:YVO_4$ laser by adopting two c-axis orthogonally a-cut bonded $YVO_4/Nd:YVO_4$ crystals. Each end face of the laser crystals was pumped by a diode laser, which allowed for the more efficient generation of orthogonally polarized linearly polarized lasers and equalized the energy of the dual-wavelength output laser [19]. In 2015, Z. Jing et al. selected the Nd:LiYF crystal as the gain medium and theoretically analyzed the conditions for simultaneous π- and σ-polarized emissions from the four-level transition in Nd^{3+}. An orthogonally polarized dual-wavelength laser at 1047, and 1053, 1321, and 1313 nm was operated in the stable CW mode by using a simple linear resonator [20]. For the 2 μm waveband, Segura et al. reported emissions operating simultaneously in the CW mode at 1922 and 1946 nm by using a $Tm:KLu(WO_4)$ crystal. The polarization switching of a $Tm:KLu(WO_4)$ laser was shown with an increased pump power [21]. X. L. Zhang et al. studied the switching of orthogonal polarizations and optical bistability in a Tm–Ho:LLF laser. The switching of the π-polarization at 2069 nm and σ-polarization at 2066 nm were experimentally observed [22]. Previous studies on dual-wavelength, orthogonally polarized lasers have mostly focused on lasers operating in the 1 μm waveband or Tm and Ho co-doped lasers in the 2 μm waveband [22–25]. However, there have been no experimental studies of the polarization characteristics of single Ho-doped LLF lasers at 2 μm.

In this paper, we demonstrate the polarization coexistence between two orthogonal polarizations of a gain-switched Ho:LLF laser. A compact intra-cavity pumped construction was adopted, and the gain-switched Ho:LLF pulse laser was obtained from a self-Q-switched Tm:YAP laser without extra Q-switched components. In this experiment, the output power of two orthogonal polarization lasers was simultaneously and individually measured. The polarization characteristics of the gain-switched Ho:LLF laser and the energy ratio between two orthogonal polarization lasers are presented.

2. Experimental Setup

For more resonant pump energy, a more compact and simple laser system with two gain mediums was used in this experiment. Figure 1 shows a simple plane-concave cavity configuration. A fiber-coupled diode laser at 795 nm was used as the pump laser. The diameter of the fiber core was 400 μm, and the numerical aperture (NA) was 0.22. The maximum pump power of the fiber-coupled diode laser was 30 W. The pump light emitted by the diode laser was linearly polarized, and the polarization direction was perpendicular to the cavity axis. To obtain a self-Q-switched laser, the diode laser pump spot was collimated and focused in the Tm:YAP crystal by a 1:2 coupling lens group with a focal length of 75 mm, and the distance between the crystal and the coupling lens group was 74 mm. Optimizing the pump spot on the gain media by slightly adjusting the position of the coupling lens set was the best way to maximize the output power. The Tm:YAP crystal and the Ho:LLF crystal were placed inside the cavity in sequence. The dimensions of the b-cut Tm:YAP crystal with a 5% dopant concentration was $3 \times 3 \times 10$ mm^3, and the a-cut Ho:LLF crystal had a dopant concentration of 0.5% with dimensions of $3 \times 3 \times 5$ mm^3. Two laser crystals were covered with indium foil and fitted on a water-cooled copper holder. The temperature of the cooling water was maintained at 291 K. The distance between the Tm:YAP crystal and the Ho:LLF crystal was 10 mm. Both end faces of the Tm:YAP crystal were high-transmission (T > 99.5%) coated at 1.9–2.1 μm and 795 nm. Regarding the Ho:LLF crystal, the dichromatic coatings on the both of its end faces were high-transmitting in the 1900–2100 nm wavelength range. A compact linear resonator was adopted between the input plane mirror and output concave mirror. The total length of the cavity was about 155 mm. The dichroic input plane mirror (M1) coating on both end faces was high-transmitting at 780–800 nm, but it was totally reflecting at 1.9–2.1 μm. A plane-concave mirror, which was partially reflective (T = 5%) at 1.9–2.1 μm, was used as the output mirror (M2). The curvature radius of the output mirror (M2) was 200 mm. A broadband polarizing beam splitter (PBS) at 1.9–2.1 μm (extinction ratio Tp:Ts > 1000:1) was placed behind the output mirror (M2) to split the output laser into two orthogonal polarizations.

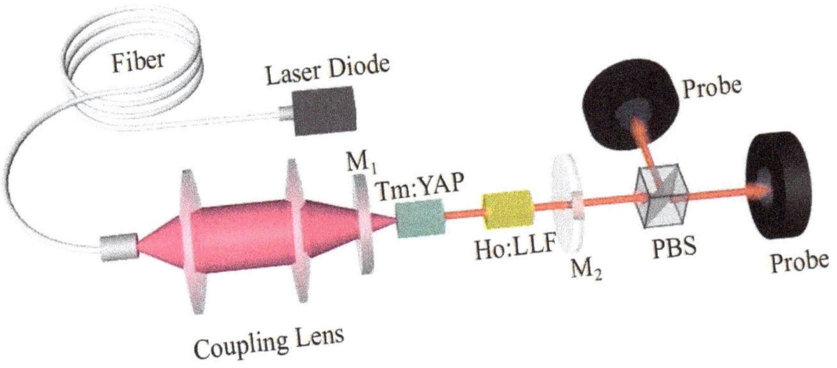

Figure 1. Experimental setup of the orthogonally polarized Ho:LLF laser.

The output power was measured with a Coherent FieldMaxII laser power meter and an InGaAs photodiode. The laser spectrum was monitored with a Zolix monochromator (Omni-λ 3015; the resolution of the spectrum was 0.1 nm). A 300 MHz bandwidth digital oscilloscope (Tektronix TDS3032B) was applied to record the output pulse trains with a >100 MHz bandwidth IR detector (Vigo PVM-10.6).

3. Experimental Results and Discussion

First, we only inserted the Tm:YAP crystal into the cavity to establish a self-Q-switched Tm:YAP pulsed laser at 1.94 μm. The self-Q-switched Tm:YAP laser operated as the second pump source to activate the Ho:LLF crystal to generate a 2 μm laser. Therefore, the Ho:LLF laser operated in the gain-switched mode. Figure 2a depicts the output power of the Tm:YAP pulse laser without the Ho:LLF crystal in the cavity as the pump power increased. Figure 2b shows pulse trains of the Tm:YAP pulse laser at a pump power of 13.5 W.

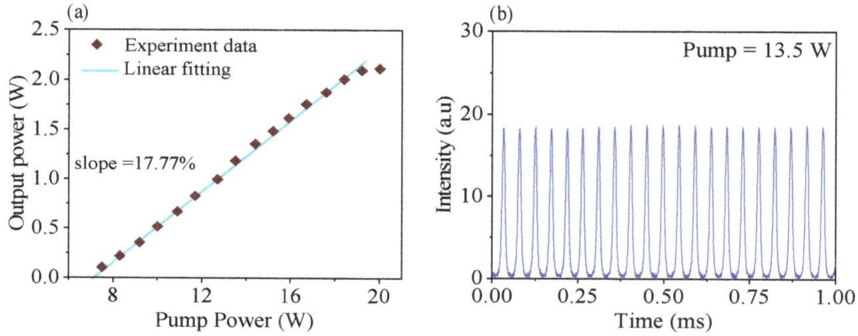

Figure 2. (**a**) Output power of the self-Q-switched Tm:YAP pulsed laser with respect to the incident pump power and (**b**) pulse trains of the Tm:YAP laser at a pump power of 13.5 W.

The maximum output power of the Tm:YAP laser at 1944 nm was 2.10 W. The slope efficiency was 17.77%. Note that the output power of the self-Q-switched Tm:YAP laser led to a saturable effect after the pump power increased to 20 W, as shown in Figure 2a. As the pump further increased above 20 W, the laser power suddenly dropped and then the laser output terminated due to the possible laser damage of the Tm:YAP crystal. This suggested that significant thermal effects occurred in the Tm:YAP crystal. The pulse laser could steadily operate for several hours. After using a narrow-band filter at 1850–1950 nm to separate the Tm:YAP laser, we measured the repetition rate and pulse width by directly collecting output signals with a digital oscilloscope. The maximum Tm:YAP pulse energy and repetition rate were 28 μJ and 47 kHz, respectively, and the minimum Tm:YAP pulse width was 2.5 μs. Because the cavity mode at 1.94 μm was larger than the 795 nm pump region, the ground state reabsorption (GSRA) effect resulted in part of the cavity mode exceeding the pump mode focused on the Tm:YAP crystal, which can accordingly be considered a saturable absorber. This provided an additional loss for the Tm:YAP laser's oscillation. The GSRA effect occurring in the Tm:YAP crystal could modulate the laser's oscillation dynamics, leading to a self-Q-switched pulse laser output [26]. The GSRA effect via Tm $^3H_6 \rightarrow {}^3F_4$ should be considered in the pumping process of lasers at 795 nm [26]. By using such a self-Q-switched laser, the Ho:LLF crystal can obtain enough intra-cavity energy to emit the laser. Figure 3 shows the energy level diagram of the Ho^{3+} $^5I_7$ excited manifold and the Ho^{3+} $^5I_8$ ground manifold of Ho:LLF. The main absorption wavelength of the Ho:LLF crystal was centered around 1940 nm, which corresponded to the emission wavelength of the Tm:YAP crystal. The large size of the Lu^{3+} host ion in Ho:LLF resulted in a larger crystal field and the crystal field splitting of the lanthanide ion [13,14]. The emission wavelengths on the π-polarization and σ-polarization in Ho:LLF were 2052 nm and 2066 nm, respectively.

After inserting the Ho:LLF crystal into the cavity, the output performance of the intra-cavity pumped Ho:LLF laser was investigated. To obtain 2 μm orthogonal polarization lasers, we set a small angle between the optical axis of the Ho:LLF laser crystal and the axis of the cavity, which could induce the high natural birefringence of the a-cut Ho:LLF crystal. To accurately measure the output profiles of the Ho:LLF laser, we used a narrow-band filter at 2000–2100 nm wavelengths to eliminate the influence of the pump laser and the Tm:YAP laser on the measurements. By using a polarizing beam splitter (PBS), the two separated orthogonally polarized laser beams were obtained. We measured the output power for individual polarization at the same time. We defined the π-polarization and the σ-polarization as the directions along the c-axis and a-axis of the Ho:LLF crystal, respectively. In order to optimize laser output quality and reduce cavity leakage loss, we compressed the length of the resonator as much as possible. Considering the actual size of the crystal and heat sink in the experiment, as well as the necessity to leave some space to rotate the Ho:LLF crystal at a small angle, the length of the resonator could be compressed to a minimum of about 155 mm. We conducted optimal output experiments by using Ho:LLF crystals of different lengths (3 mm, 5 mm and 8 mm). Figure 4a shows the variation of output power with pump power for the three different crystal lengths. As shown in Figure 4a, the best result was obtained with the Ho:LLF crystal dimensions of $3 \times 3 \times 5$ mm^3. The total output power and the output power for the individual polarization directions were measured by increasing the pump power. While increasing the pump power from zero to threshold power at $P_{\pi\text{-thr}}$ = 13.5 W, the output power varied from zero to 26.7 mW. Figure 4b,c, respectively, shows the total output power and the output power for the individual polarization directions of the Ho:LLF (crystal length of 5 mm) laser with respect to the incident pump power. When pump power increased above the threshold, the output laser was only π-polarized at 2052 nm, and the maximum output power was 299 mW when the pump power was 15.9 W. After increasing pump power from $P_{\pi\text{-thr}}$ to $P_{\pi\text{-max}}$ = 15.9 W, the output power for π-polarization in the Ho:LLF laser decreased. When the pump power increased from $P_{\sigma\text{-thr}}$ = 16.3 W to 18.8 W, the σ-polarization laser at 2066 nm began oscillating and the output power for σ-polarization increased from zero to maximum power $P_{\sigma\text{-max}}$ = 126 mW. By inserting two laser crystals into the cavity, the laser of the Tm:YAP crystal could not only be used as a pump source to provide gain to the Ho:LLF crystal but also control the losses of the orthogonal polarizations in the Ho:LLF laser based on the excellent anisotropy of the Tm:YAP crystal. Therefore, the variation of the output power of the two orthogonally polarized laser could be controlled by changing the energy of the Tm:YAP laser. Because of the tiny angle between the optical axis of Ho:LLF and the axis of the cavity, the Ho:LLF laser could emit two orthogonally polarized lasers based on the high natural birefringence of the Ho:LLF crystal [7,25]. In addition, the $^5I_7 \rightarrow {}^5I_8$ energy level transition resulted in 2052 and 2066 nm laser lines. These two laser lines shared a common upper level. The gain competition between 2052 and 2066 nm is not negligible, especially when the pumping power is further increased [23]. Figure 4c shows that when increasing the pump power from $P_{\sigma\text{-thr}}$, the output laser energy for the π- and σ-polarizations tended to enter the equilibrium zone. The output power of the π-polarization (2052 nm) was almost the same as the output power of the σ-polarization (2066 nm) when the pump power was 18.4 W.

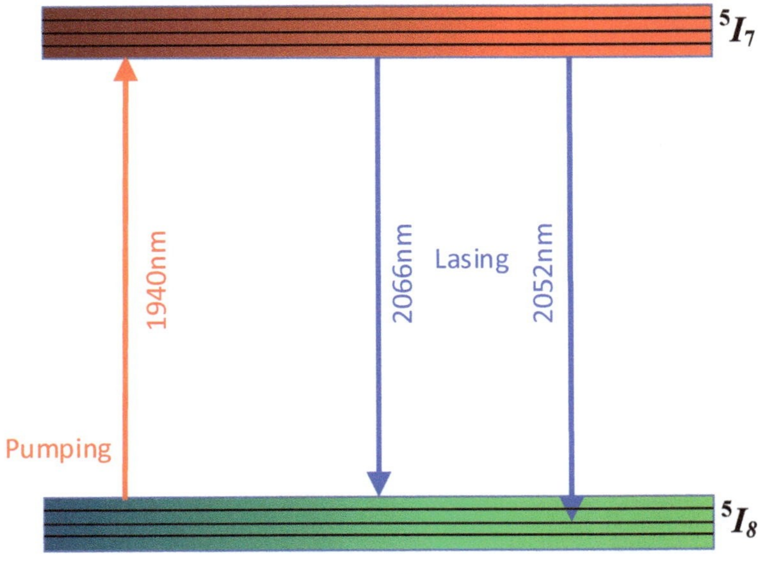

Figure 3. Energy levels in the Ho:LLF crystal [13,14].

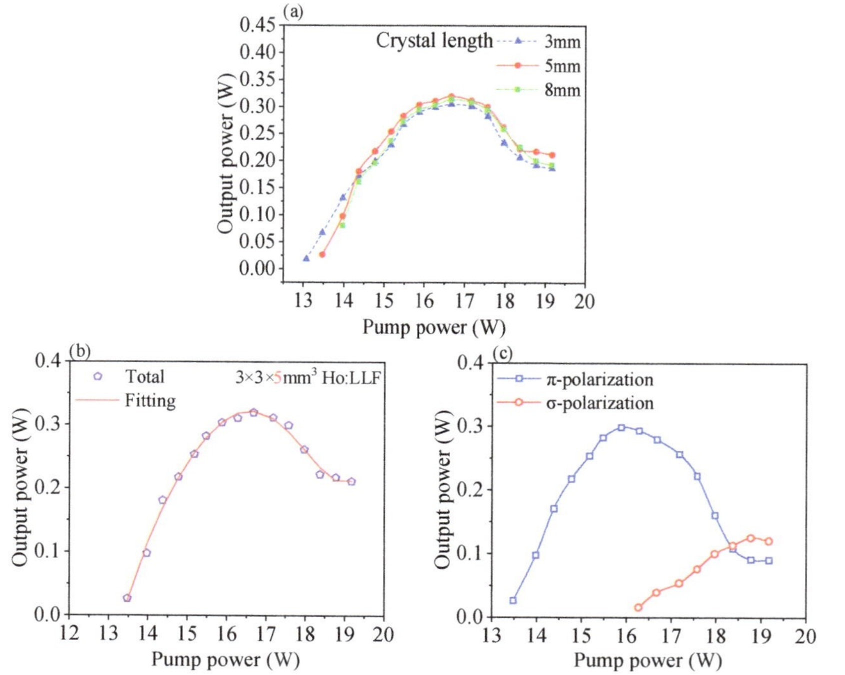

Figure 4. (**a**) The output power and pump power for different crystal lengths (3 mm, 5 mm, and 8 mm); (**b**) the total and (**c**) individual polarization direction output power in the Ho:LLF (crystal length was 5 mm) laser as a function of the incident pump power.

The measured spectra for the π-polarization and the σ-polarization of the Ho:LLF laser are shown in Figure 5a. The central wavelengths were 2052 nm and 2066 nm for the orthogonal π- and σ-polarizations, respectively. The variation of the emission wavelength for the gain-switched Ho:LLF laser can be explained in the process of intra-cavity gain variation by increasing the pump power, as shown in Figure 5b. With increases in the pump power, the 2052 nm and 2066 nm spectral widths for the π- and σ-polarization emission laser lines, respectively, began to increase due to decreases in the pulse width [27]. The full width at half maximum (FWHM) of the 2052 nm and 2066 nm wavelengths for the orthogonal polarizations were 3.2 nm and 2.6 nm, respectively, as shown in Figure 5a.

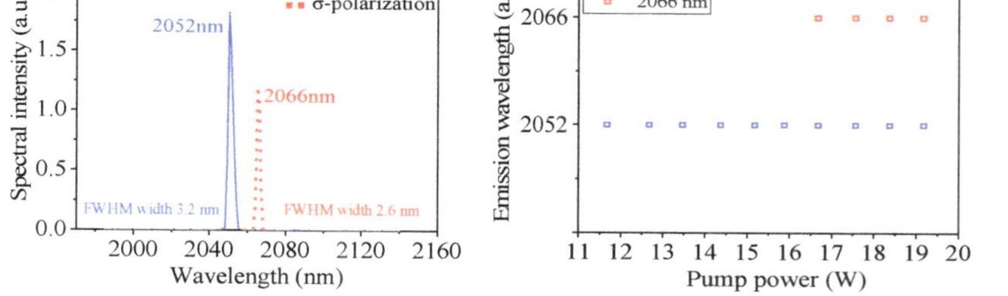

Figure 5. (**a**) Optical spectra of the orthogonally polarized dual-wavelength Ho:LLF laser and (**b**) emission wavelengths of the Ho:LLF laser with respect to pump power.

In addition, the linearly polarized pump induced the parallel polarization of the self-Q-switched Tm:YAP laser. Then, the polarized Tm:YAP laser's intra-cavity pumped the

Ho:LLF crystal to produce a parallel π-polarization and a vertical σ-polarization at dual wavelengths of 2052 nm and 2066 nm. The larger stimulated emission cross-section of the π-polarization (along the c-axis) compared with that of the σ-polarization (along the a-axis) [18] resulted in different oscillation thresholds at the two orthogonal polarizations and different output energy levels at the dual wavelengths, as shown in Figure 4.

When the pump power was maintained 18.4 W, we rotated a linear polarizer to observe the relation between the angle of polarization and the output power after splitting with the PBS, as shown in Figure 6a. The output energy of the π-polarization and σ-polarization lasers was almost the same as when the pump power was 18.4 W. The angle difference of the output power for the π-polarization and σ-polarization directions was 90 degrees, which means that the polarization direction was mutually orthogonal. The process of energy changes within the laser system allowed the entire laser system to achieve the co-existence of two orthogonally polarized lasers due to gain variation by controlling the pump power.

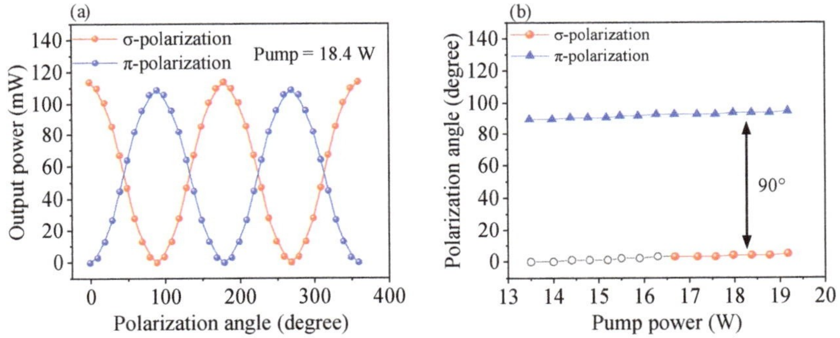

Figure 6. (**a**) The relationship between the angle of polarization and the output power; (**b**) polarization angles of the orthogonally polarized laser as a function of the pump power.

The variation of the polarization directions in the Ho:LLF laser caused by increases in the pump power is shown in Figure 6b. Because of the high natural birefringence of the a-cut Ho:LLF crystal, the rotation of the polarization directions of the output power did not occur when the pump power increased [22,23]. Figure 6b shows that the two polarization directions remained mutually orthogonal at all times.

Furthermore, due to the gain-switch between the dual crystals placed into the cavity, the output laser of the Ho:LLF crystal was a pulse laser. Figure 7a,b shows the pulse trains and single pulse shape of the gain-switched Ho:LLF laser when the pump power was 19.2 W. The gain-switched Ho:LLF pulse laser at the 2 μm waveband was only obtained with the self-Q-switched Tm:YAP laser. The intensity and period of the gain-switched Ho:LLF laser were stable. The repetition rate and the pulse width of the gain-switched Ho:LLF pulsed laser with respect to the pump power are shown in Figure 8a,b. When increasing the pump power from 13.5 W to 19.2 W, the output pulse energy was insensitive and remained nearly constant at 28 μJ. The repetition rate gradually increased to the maximum of 24 kHz when the pump power increased to 19.2 W. The minimum pulse width was measured as 326 ns. This increasing tendency occurred because as the pump power increased, the rate of the population inversion density on the upper-level Ho rapidly accumulated and the time required for the population inversion density to reach the threshold was shortened. This led to an increase in the pulse repetition rate and a parallel reduction in the pulse width [28,29]. Due to mechanical and thermal perturbation, the output pulse profile somewhat fluctuated. In this experiment, we employed an air-bearing vibration isolation platform to suppress the fluctuation of the output pulse profiles caused by mechanical vibration. This effectively reduced the uncertainty in the measurements.

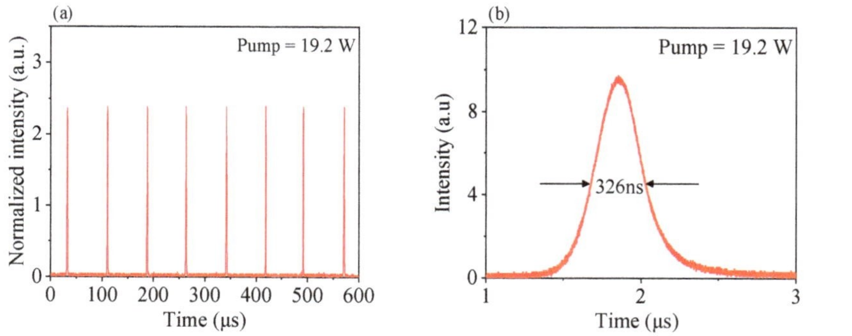

Figure 7. (**a**) Pulse trains and (**b**) single pulse shape of the gain-switched Ho:LLF laser at a pump power of 19.2 W.

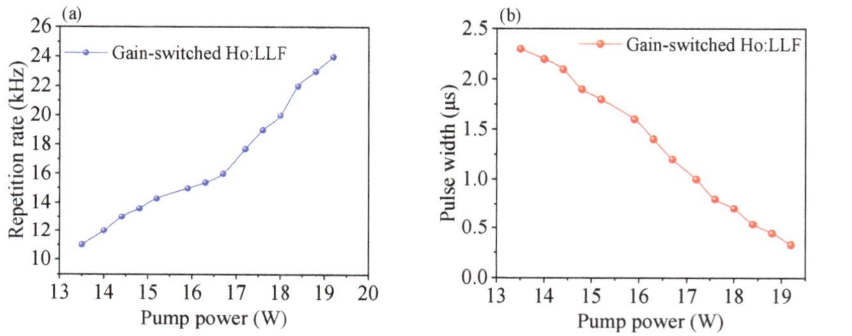

Figure 8. (**a**) Repetition rate and (**b**) temporal profile of the gain-switched Ho:LLF laser as a function of pump power.

4. Conclusions

In conclusion, we experimentally demonstrated output parameters for π- and σ-polarizations at two different wavelengths in a gain-switched Ho:LLF laser by using an intra-cavity self-Q-switched Tm:YAP laser. The co-existence of the σ-polarized laser at 2066 nm and the π-polarized laser at 2052 nm was achieved by increasing the pump power. The maximum output power for π-polarization (2052 nm) and σ-polarization (2066 nm) was 299 mW and 126 mW at pump power values of 15.9 W and 18.8 W, respectively. The output energy of the two orthogonally polarized states could reach a balanced region in which the output power of the π- and σ-polarization lasers was almost the same when the pump power increased from 18.0 W to 19.2 W. The balanced output energy of the orthogonally polarized dual-wavelength laser could be widely used in differential absorption radars, and the precise measurement of CO_2 and H_2O in the atmosphere can be achieved via loss-free switching between the dual-wavelength lasers. Regarding the scaling of the output energies of the dual-wavelength laser pulses generated by the Ho:LLF laser, one can make efforts to improve the thermal effects of the Tm:YAP crystal to enhance the intra-cavity pumping power. After measuring the variation of the polarization angle and the increase in pump power, it was found that the two polarization directions were stable and remained orthogonal to each other at all times because of the high natural birefringence of the a-cut Ho:LLF crystal. By using a PBS and other polarized components, we could achieve the switching of the σ-polarization at 2066 nm and the π-polarization at 2052 nm. Furthermore, the minimum pulse width of the gain-switched Ho:LLF laser was about 326 ns, and the maximum pulse energy was 28 μJ. The pulse repetition rate could be tuned from 11 kHz to 24 kHz by adjusting the pump power. The orthogonally polarized dual-wavelength gain-switched Ho:$LuLiF_4$ laser has potential applications in the generation of terahertz waves, as a light source of fiber lasers, and in differential absorption lidars.

Author Contributions: Conceptualization, G.H., X.Z. and L.L.; data curation, G.H., Z.G., X.Z. and L.L.; investigation, J.C., X.Z., F.T., L.L. and G.H.; writing—review and editing, G.H., Z.G., J.C., S.Y., S.L., F.T., X.Z. and L.L.; funding acquisition, J.C., F.T., X.Z. and L.L.; formal analysis, L.L.; writing—original draft preparation, G.H.; project administration, J.C. and X.Z.; supervision, L.L. All authors have read and agreed to the published version of the manuscript.

Funding: This research was funded by the National Natural Science Foundation of China, grant numbers 61875043 and 72275194, and the 111 Project to the Harbin Engineering University, grant number B13015.

Institutional Review Board Statement: Not applicable.

Informed Consent Statement: Not applicable.

Data Availability Statement: Data underlying the results presented in this paper are not publicly available at this time but may be obtained from the authors upon request.

Conflicts of Interest: The authors declare no conflict of interest.

References

1. Yu, J.; Trieu, B.C.; Modlin, E.A.; Singh, U.N.; Kavaya, M.J.; Chen, S.; Bai, Y.; Petzar, P.J.; Petros, M. 1 J/pulse Q-switched 2 microm solid-state laser. *Opt. Lett.* **2006**, *31*, 462–464. [CrossRef]
2. Taczak, T.M.; Killinger, D.K. Development of a tunable, narrow-linewidth, cw 2.066-μm Ho:YLF laser for remote sensing of atmospheric CO_2 and H_2O. *Appl. Opt.* **1998**, *37*, 8460–8476. [CrossRef]
3. Budni, P.A.; Pomeranz, L.A.; Lemons, M.L.; Miller, C.A.; Mosto, J.R.; Chicklis, E.P. Efficient mid-infrared laser using 1.9-μm-pumped Ho:YAG and $ZnGeP_2$ optical parametric oscillators. *J. Opt. Soc. Am. B* **2000**, *17*, 723–728. [CrossRef]
4. Scholle, K.; Lamrini, S.; Koopmann, P.; Fuhrberg, P. 2 μm laser sources and their possible applications. In *Frontiers in Guided Wave Optics and Optoelectronics*; Bishnu, P., Ed.; Intech: London, UK, 2010.
5. He, C.; Killinger, D.K. Dual-polarization modes and self-heterodyne noise in a single-frequency 2.1-microm microchip Ho,Tm:YAG laser. *Opt. Lett.* **1994**, *19*, 396–398. [CrossRef] [PubMed]
6. Qi, Y.; Yang, S.; Wang, J.; Li, L.; Bai, Z.; Wang, Y.; Lv, Z. Recent advance of emerging low-dimensional materials for vector soliton generation in fiber lasers. *Mater. Today Phys.* **2022**, *23*, 100622. [CrossRef]
7. Schellhorn, M. A comparison of resonantly pumped Ho:YLF and Ho:LLF lasers in CW and Q-switched operation under identical pump conditions. *Appl. Phys. B* **2011**, *103*, 777–788. [CrossRef]
8. Na, Q.X.; Gao, C.Q.; Wang, Q.; Zhang, Y.X.; Gao, M.W.; Ye, Q.; Li, Y. 15 mJ single-frequency Ho:YAG laser resonantly pumped by a 1.9 μm laser diod. *Laser Phys. Lett.* **2016**, *13*, 095003. [CrossRef]
9. Tang, Y.L.; Xu, L.; Wang, M.J.; Yang, Y.; Xu, X.D.; Xu, J.Q. High-Power gain-switched Ho:LuAG rod laser. *Laser Phys. Lett.* **2010**, *8*, 120–124. [CrossRef]
10. Chen, H.; Shen, D.; Zhang, J.; Yang, H.; Tang, D.; Zhao, T.; Yang, X. In-band pumped highly efficient Ho:YAG ceramic laser with 21 W output power at 2097 nm. *Opt. Lett.* **2011**, *36*, 1575–1577. [CrossRef] [PubMed]
11. Chao, C.; Ju, Y.L.; Yao, B.Q.; Dai, T.Y.; Duan, X.M.; Li, J.; Ding, Y.; Liu, W.; Pan, Y.B.; Li, C.Y. Passively Q-switched Ho:YLF laser pumped by Tm^{3+} -doped fiber laser. *Opt. Laser Technol.* **2016**, *77*, 55–58.
12. Walsh, B.M.; Barnes, N.P.; Petros, M.; Yu, J.; Singh, U.N. Spectroscopy and modeling of solid state lanthanide lasers: Application to trivalent Tm^{3+} and Ho^{3+} in $YLiF_4$ and $LuLiF_4$. *J. Appl. Phys.* **2004**, *95*, 3255–3271. [CrossRef]
13. Zhao, C.; Hang, Y.; Zhang, L.; Yin, J.; Hu, P.; Ma, E. Polarized spectroscopic properties of Ho^{3+}-doped $LuLiF_4$ single crystal for 2 μm and 2.9 μm lasers. *Opt. Mater.* **2011**, *33*, 1610–1615. [CrossRef]
14. Walsh, B.M.; Grew, G.W.; Barnes, N.P. Energy levels and intensity parameters of Ho^{3+} ions in $GdLiF_4$, $YLiF_4$ and $LuLiF_4$. *J. Phys. Condens. Matter* **2005**, *17*, 7643. [CrossRef]
15. Schellhorn, M. High-energy, in-band pumped Q-switched Ho^{3+}:$LuLiF_4$ 2 μm laser. *Opt. Lett.* **2010**, *35*, 2609–2611. [CrossRef]
16. Schellhorn, M. High-energy, in-band pumped Ho:LLF MOPA system. In *Advanced Solid-State Photonics*; Optical Society of America: Washington, DC, USA, 2012; p. AW4A-4.
17. Walsh, B.M. Dual wavelength lasers. *Laser Phys.* **2010**, *20*, 622–634. [CrossRef]
18. Vodopyanov, K.L.; Fejer, M.M.; Simanovskii, D.M.; Kozlov, V.G.; Lee, Y.S. Terahertz-wave generation in periodically-inverted GaAs. In Proceedings of the Conference on Lasers and Electro-Optics, Baltimore, MD, USA, 22–27 May 2005; p. CWM1.
19. Xing-Peng, Y.; Qiang, L.; Hai-Long, C.; Xing, F.; Ma-Li, G.; Dong-Sheng, W. A novel orthogonally linearly polarized Nd:YVO_4 laser. *Chin. Phys. B* **2010**, *19*, 395–400. [CrossRef]
20. Jing Zhang, J.Z.; HuiLong Li, H.L.; Jing Xia, J.X.; Xihong Fu, X.F. Orthogonally polarized dual-wavelength Nd:$YLiF_4$ laser. *Chin. Opt. Lett.* **2015**, *13*, 37–42. [CrossRef]

21. Segura, M.; Kadankov, M.; Mateos, X.; Pujol, M.C.; Carvajal, J.J.; Aguiló, M.; Díaz, F.; Griebner, U.; Petrov, V. Polarization switching in the 2-m Tm:KLu$(WO_4)_2$ laser. *Laser Phys. Lett.* **2012**, *9*, 104–109. [CrossRef]
22. Zhang, X.L.; Zhang, S.; Xie, Z.L.; Li, G.X.; Yu, L.; Cui, J.H.; Zhao, J.Q.; Li, L. Polarization switching and optical bistability in the diode-pumped Tm,Ho:LLF laser. *Laser Phys. Lett.* **2014**, *11*, 1528–1529. [CrossRef]
23. Zhang, X.; Zhang, S.; Wang, C.; Li, L.; Zhao, J.; Cui, J. Orthogonally polarized dual-wavelength single-longitudinal-mode Tm,Ho:LLF laser. *Opt. Express* **2013**, *21*, 22699–22704. [CrossRef]
24. Zhang, X.L.; Yu, L.; Zhang, S.; Li, L.; Zhao, J.Q.; Cui, J.H.; Dong, G.Z.; Wang, R. Controlled optical bistability switching in a diode-pumped Tm,Ho:LLF laser. *Laser Phys. Lett.* **2013**, *10*, 5801. [CrossRef]
25. Barnes, N.P.; Walsh, B.M.; Filer, E.D. Ho:Ho upconversion: Applications to Ho lasers. *J. Opt. Soc. Am. B* **2003**, *20*, 1212–1219. [CrossRef]
26. Kang, P.; Zhang, S.; Zhang, X.; Huang, J. High power self-Q-switched Tm:YAP solid state laser based on the ground state reabsorption. *Infrared Phys. Technol.* **2022**, *123*, 104117. [CrossRef]
27. Liu, H.; Cao, S.; Wang, W.; Lin, B.; Lu, W.; Fang, Z. A 1μm laser output based on an Er-doped fiber femtosecond laser. *Opt. Commun.* **2017**, *397*, 161–165. [CrossRef]
28. Wang, M.Q.; Ding, S.H.; Yu, W.Y.; Zhang, W.H. High-efficient diode-pumped passively Q-switched c-cut Nd:GdVO4self-Raman laser. *Laser Phys. Lett.* **2013**, *10*, 045403. [CrossRef]
29. Liu, J.; Yang, J.; He, J. High repetition frequency passively Q-switched diode-pumped Nd:YVO4 laser. *Opt. Laser Technol.* **2003**, *35*, 431–434. [CrossRef]

photonics

Communication

Effect of Contact Angle on Friction Properties of Superhydrophobic Nickel Surface

Junyuan Huang [1,†], Zhiwei Zhu [2,†], Ling Zhang [3,4,*], Dongdong Guo [5], Zhen Niu [5] and Wei Zhang [1,*]

1 School of Mechanical Engineering and Automation, Beihang University, Beijing 100191, China; jyhuang@buaa.edu.cn
2 Department of Mechanical and Automation Engineering, The Chinese University of Hong Kong, Hong Kong, China; zhiweizhu@cuhk.edu.hk
3 Key Lab of Semiconductor Materials Science, Institute of Semiconductors, Chinese Academy of Sciences, Beijing 100083, China
4 Joint Lab of Digital Optical Chip, Wuyi University, Jiangmen 529020, China
5 MRA2 Bodyshop, Beijing Benz Automotive Co., Ltd., Beijing 100176, China; guodd@bbac.com.cn (D.G.); niuzhen@bbac.com.cn (Z.N.)
* Correspondence: zhangling@semi.ac.cn (L.Z.); zhangweibh@buaa.edu.cn (W.Z.)
† These authors contributed equally to this work.

Abstract: Due to their excellent performance, superhydrophobic materials have received a lot of attention and research in friction reduction and wear resistance. However, the effect of different contact angles of superhydrophobicity on friction and wear properties has not been thoroughly studied. In this paper, a nanosecond pulsed laser was used to realize the preparation of a superhydrophobic nickel surface, which indeed reduced the coefficient of friction but also increased the wear volume when compared to the unprocessed surface. As the contact angle of the superhydrophobic nickel surface increased, the coefficient of friction gradually increased, and the wear volume decreased gradually in superhydrophobic nickel surfaces. When the laser energy density was 1 J/cm^2, the contact angle was 150.3° and the minimum friction coefficient was 0.4. However, when the laser energy density was 50 J/cm^2, the maximum contact angle was 156.4° and the minimum wear volume was 4.23×10^7 μm^3. The friction direction also influenced the tribological properties of the superhydrophobic-textured surface. This method makes it possible to process superhydrophobic surfaces with more suitable friction and wear properties.

Keywords: laser surface texturing; nickel; superhydrophobicity; tribology; contact angle

Citation: Huang, J.; Zhu, Z.; Zhang, L.; Guo, D.; Niu, Z.; Zhang, W. Effect of Contact Angle on Friction Properties of Superhydrophobic Nickel Surface. *Photonics* **2023**, *10*, 829. https://doi.org/10.3390/photonics10070829

Received: 20 June 2023
Revised: 6 July 2023
Accepted: 14 July 2023
Published: 17 July 2023

1. Introduction

Wettability is an important characteristic of solid surfaces, and the contact angle is the most used parameter to characterize wettability [1]. The material is superhydrophobic when the water droplet contact angle is greater than 150° and the water droplet sliding angle is less than or equal to 10° [2]. Studies have shown that hydrophobicity can change the properties of the material surface by changing the chemical composition of the material surface and the geometry of the micro/nanostructure, making superhydrophobic materials useful [3,4].

At present, methods for preparing superhydrophobic surfaces on metals include hydrothermal [5], chemical deposition [6], electrodeposition [7], etc. Compared with other superhydrophobic surface fabrication methods, the operation process of laser surface texturing is simple, non-toxic, non-polluting, and has high stability and diversity of processed materials [8,9]. Many studies have discussed laser preparation of superhydrophobic metal surfaces. Rajab [10] obtained a stable superhydrophobic surface on 316 L stainless steel using laser surface texturing. Huang [11] fabricated a superhydrophobic surface on aluminum alloy using laser treatment coupled with a chemical functionalization method.

Lu [12] illustrated that the laser-textured superhydrophobic surface enhanced the corrosion resistance dramatically. Shimada [13] proposed a concept of direct laser processing of two-scale periodic structures exhibiting superhydrophobicity, with the maximum apparent contact angle of 161.4° and the contact angle hysteresis of 4.2° for a pitch of 80 μm and 20 repetition shots. Huang [14] used an ultrafast femtosecond laser to prepare a superhydrophobic/superlipophilic lead bronze surface and found that the anisotropy of the surface structure led to the anisotropy of the tribological properties.

Due to their excellent performance, superhydrophobic materials have good performance in anti-friction [15], anti-corrosion [16], antibacterial [17], etc., and are widely used in aerospace, petroleum engineering, and other fields. Among them, anti-friction and anti-wear have received a lot of attention and research. Jiao [18] manufactured a superhydrophobic aluminum surface with great wear-resistance and decreased the friction force of the sliding interface. Guo [19] carried out nanosecond laser processing to construct a grid of microgrooves on an aluminum surface and reported whether it was under dry friction or water lubrication conditions, with the obtained superhydrophobic samples resulting in good anti-friction and wear-resistant ability. However, the influence of different contact angles of superhydrophobicity on the friction and wear properties has not been thoroughly studied. It is still unknown how large the contact angle of the metal surface can be prepared to have the best tribological properties.

In this paper, the common nickel was used to prepare a nickel surface with superhydrophobic properties by using nanosecond laser processing technology. The effect of laser energy density on wettability was studied, and the friction and wear properties of superhydrophobic surfaces with different contact angles were compared and analyzed. We also analyzed the effect of perpendicular and parallel friction directions on the tribological behavior of superhydrophobic surfaces. This is very important for producing superhydrophobic surfaces with specified tribo-testing directions, which is beneficial for the development of superhydrophobic surfaces with the best contact angle.

2. Experimental Details

2.1. Sample Preparation

The pure nickel content in the nickel material was greater than 99.5%, which was grinded by 800#, 1000#, and 1500# sandpaper successively to ensure that the roughness of the sample was consistent and smooth before the experiment. And then they were ultrasonic treated with acetone, anhydrous ethanol, and deionized water, respectively, for 5 min and dried with compressed air to remove the impurities on the surfaces. In addition, the above ultrasonic cleaning process was still required after the laser surface texturing for the samples, as shown in Figure 1. The wettability and tribology properties were measured after the samples were exposed to the air for two weeks. In our previous experiments [20], most of the samples achieved superhydrophobicity by exposing nickel to air for two weeks. Of course, the longer the time, the better the superhydrophobicity may be. However, this experiment did not focus on the effect of time on the superhydrophobic properties of the surface, so the performance of the surface of the sample placed at other times was not discussed.

2.2. Laser Surface Texturing

A nanosecond pulsed fiber laser (Sanda laser, YLP-SD20L, Wuhan, China) with 1064 nm center wavelength, 20 kHz repetition frequency, and 10 W maximum average power was adopted in our experiment. The scanning galvanometer with a 110 mm focal length controlled the movement of the laser in the plane, and the focused spot diameter was less than 20 μm. The laser processing was carried out in the air with a lateral stripe scanning path. The energy density range could be set from 0 to 100 J/cm^2 with a scanning interval of 0.5 mm and a scanning speed of 20 mm/s. The experimental principle is shown in Figure 2.

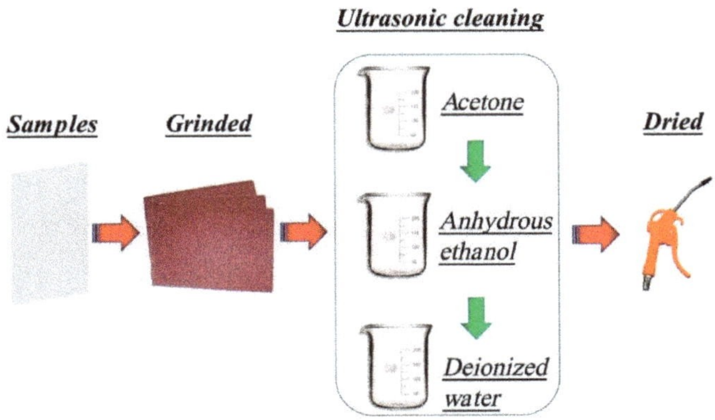

Figure 1. Nickle sample preparation processes.

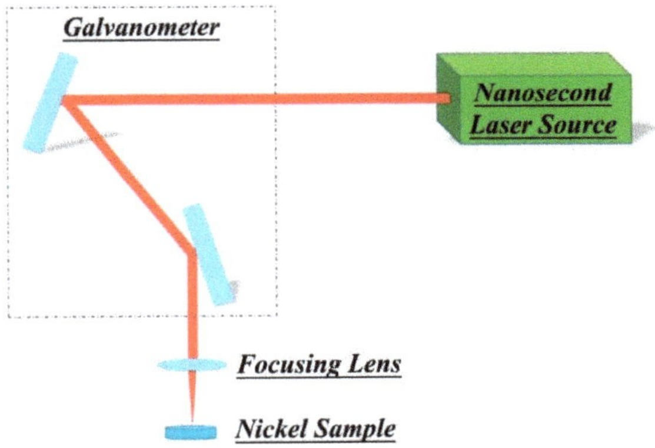

Figure 2. Experimental schematic.

2.3. The Measurement of the Wettability

The water contact angle of the surfaces was measured by the contact angle measuring instrument (OCA20). The surface topography of the samples was observed by the scanning electron microscope (SEM) and the ZYGONexView white light interferometer (WLI).

2.4. The Antifriction Performance

The SRV-4 friction and wear testing machine was used to analyze the antifriction performance for the point of contact. The friction parts were Si_3N_4 ceramics. The parameters were load of 10 N, time of 20 min, stroke of 1000 mm, and frequency of 20 kHz under dry friction conditions, and each test was repeated three times.

3. Results

3.1. Effect of Laser Energy Density on Contact Angle

According to various laser energy densities, the nickel surface was processed, and the measured contact angle is shown in Figure 3. When the energy density was less than 1 J/cm^2, the contact angle increased almost linearly with the increase in energy density, from 60.8° on the unprocessed surface to 144.7° at 0.5 J/cm^2. At this time, the surface had strong adhesion, the surface of the sample was turned over, and the water droplets would not slip off. When the energy density was greater than 1 J/cm^2, the contact angle showed a stable trend without an obvious increase, and all of them were greater than 150°. The sample was placed on a platform with an inclination angle of 10°, and the water droplets fell off. Therefore, the samples all exhibited superhydrophobic properties.

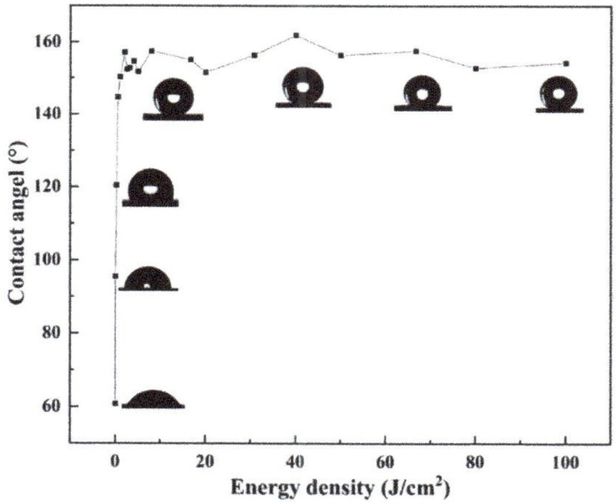

Figure 3. The variation in the water contact angle of the nickel surfaces with the laser energy density.

3.2. *Analysis of Wettability Transition Reasons*

The surface wettability of materials was mainly related to the surface roughness and surface chemical composition, so the characterization and analysis were carried out for these two aspects. In addition to the unprocessed surface, superhydrophobic samples with laser energy densities of 1, 20, 50, and 100 J/cm^2 were selected as examples and named U, L1, L20, L50, and L100, respectively.

Figure 4a–e show the SEM two-dimensional topography of samples U, L1, L20, L50, and L100. The two-dimensional diagram showed that the unprocessed surface was very smooth, and the processed superhydrophobic surface formed a periodic stripe structure. Due to the action of the laser, the material in the illuminated area was ablated and removed and then re-solidified in the non-irradiated area. Protrusions were formed on the sample surface due to the rapid temperature drop caused by the movement of the laser beam. These protrusions were attached to a smaller cylindrical protrusion that was spherical and disc-shaped, ranging from a few micrometers to ten micrometers. The striped-textured surface presents a multi-scale micro-nano structure. The formation of these dimples and protrusions increased the surface roughness of the samples [21].

Figure 4f–j show the LCM three-dimensional topography of samples U, L1, L20, L50, and L100. The three-dimensional map and roughness results of the surface roughness after laser processing was much higher than that of the unprocessed surface, and the surface roughness gradually increased with the increase in energy density. According to the roughness results, its relationship with the contact angle was analyzed. The roughness of the superhydrophobic surface was greater than 3564 μm. The surface roughness increased, and the samples also maintain superhydrophobic properties. Therefore, the surface roughness of the samples played an important role in the superhydrophobicity.

Next, the influence of the surface chemical composition on the contact angle was analyzed. The EDS element content map of samples U, L1, L20, L50, and L100 is shown in Figure 5. Since the C content was lower than 1%, the influence of C was ignored. It was considered that the surface contains only O and Ni elements before and after laser processing. The atomic content ratio of each sample is shown in Figure 6. With the increase in the laser energy density, the atomic proportion of O on the surface gradually increased, and the increase rate also gradually decreased, and the change rule was the same as that of the roughness. The change law of the contact angle with the O atom ratio was also similar. An increase in O content represents an increase in the content of metal oxides formed, and the chemical bonds of metal oxides are polar chemical bonds [22]. The higher the content of polar chemical bonds, the higher the surface free energy, which is not conducive to the acquisition of superhydrophobicity. Considering the surface chemical composition, the increase in laser energy density was not conducive to the formation of surface superhydrophobicity.

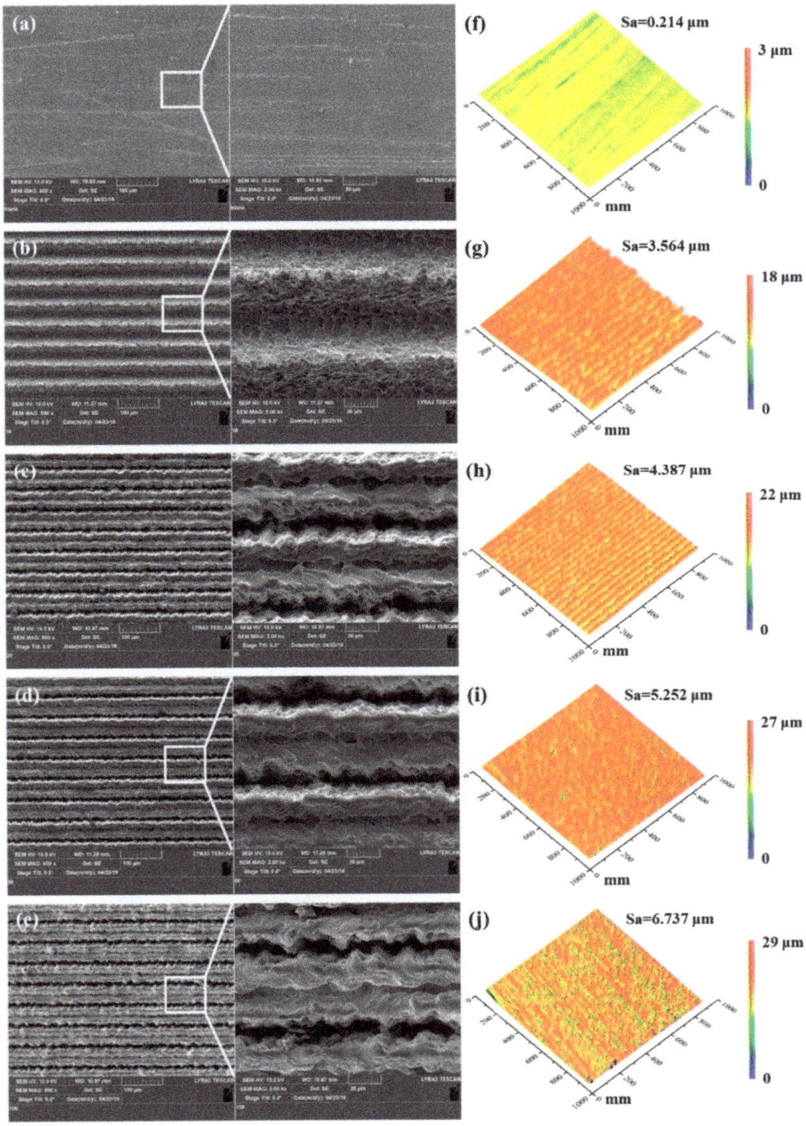

Figure 4. SEM two-dimensional topography and LCM three-dimensional topography of (**a**) U, (**b**) L1, (**c**) L20, (**d**) L50, and (**e**) L100. LCM three-dimensional topography of (**f**) U, (**g**) L1, (**h**) L20, (**i**) L50, and (**j**) L100.

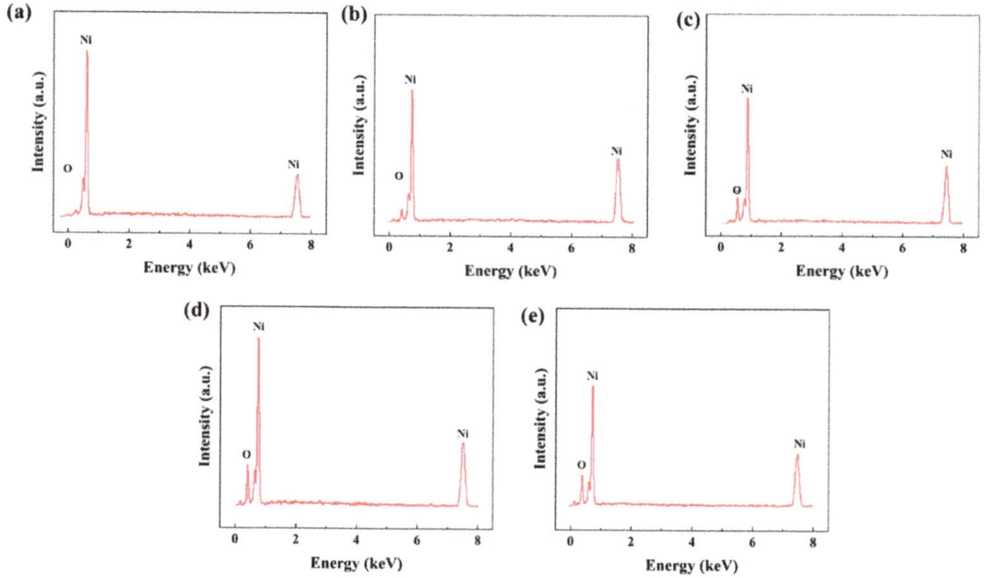

Figure 5. EDS element content diagrams of (**a**) U, (**b**) L1, (**c**) L20, (**d**) L50, and (**e**) L100.

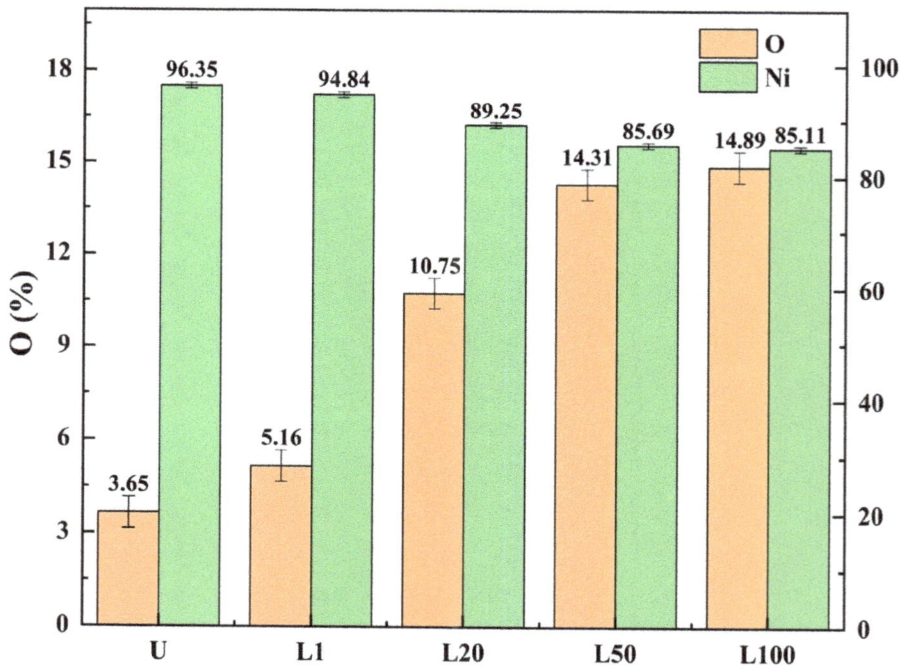

Figure 6. Atomic content ratio of U, L1, L20, L50, and L100.

The above analysis showed that the laser energy density would increase the roughness and O content of the nickel surface at the same time. Under the combined action of the two, the surface with a laser energy density greater than 1 J/cm^2 exhibited superhydrophobic properties. As the energy density increased, the surface roughness increased, which was conducive to the formation of superhydrophobic surfaces. However, currently, the proportion of O atoms was also increased, and the surface free energy was increased, which was not conducive to the formation of a superhydrophobic surface [23]. Surface roughness played a dominant role at this time, offsetting the adverse effects of surface free energy so that the surface still exhibited superhydrophobic properties. The combined effect of the two also led to the fact that the contact angle did not change monotonously with the increase in surface roughness.

4. Discussion

Next, the effect of laser energy density on the friction and wear properties of superhydrophobic surfaces was analyzed. The superhydrophobic samples with laser energy densities of 1, 20, 50, and 100 J/cm^2 were still selected for friction and wear experiments. The measured coefficient of friction (COF) curve is shown in Figure 7a. The COF of the superhydrophobic surface was reduced compared with 0.821 of sample U. The relationship between the surface contact angle and the COF was analyzed, as shown in Figure 7b. As the contact angle increased, the COF increased gradually. When the laser energy was 1 J/cm^2, the contact angle was 150.3° and the minimum friction coefficient was 0.4.

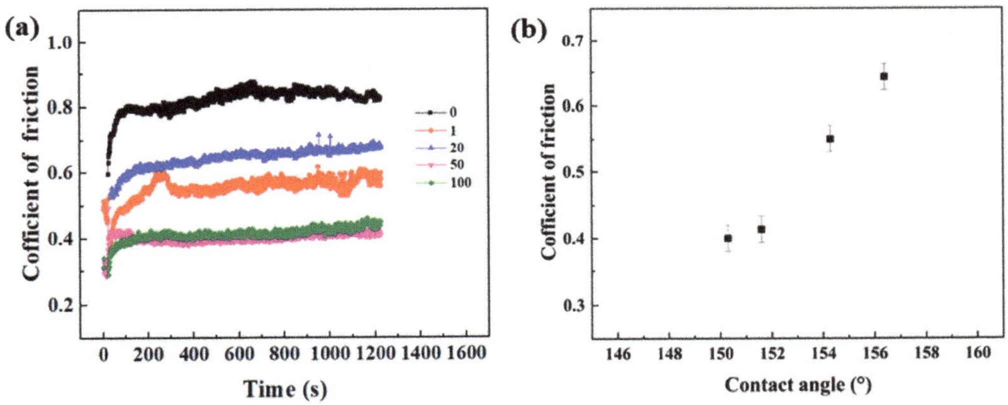

Figure 7. (**a**) Coefficient of friction (COF) curve. (**b**) Relationship between the surface contact angle and the COF.

The three-dimensional morphology of the wear scar on the surface after friction is shown in Figure 8. The width of the wear scar on the superhydrophobic surface after laser processing was larger than that of the unprocessed sample U, and the measured wear volume of sample U was 1.25×10^7 μm^3. The relationship between surface contact angle and wear volume is shown in Figure 9. The wear volume of the superhydrophobic sample surface was larger than that of sample U without processing. With the increase in the contact angle, the wear volume presented a gradually decreasing trend. At 50 J/cm^2, the maximum contact angle was 156.4° and the minimum wear volume was 4.23×10^7 μm^3.

The laser-fabricated superhydrophobic surface reduced the COF of nickel but also increased the wear volume. The superhydrophobic property of the surface reduced the adhesion of the surface of the friction pair and reduced the COF. Due to the thermal effect of laser processing, the hardness of the surface micro-nano texture was greater than that of the metal itself. High-hardness abrasive debris would participate in the subsequent friction and wear process, and the resulting ditch effect would increase the wear volume [24]. However, as the contact angle increased, the removal effect of the laser on the material surface increased, the removed surface area increased, and the contact area with the friction piece decreased; therefore, the COF gradually increased [25]. At the same time, when the contact angle increased, the surface roughness also increased. Currently, the surface micro-nano structure increased, and the surface captured the wear debris, which led to a gradual decrease in the wear volume.

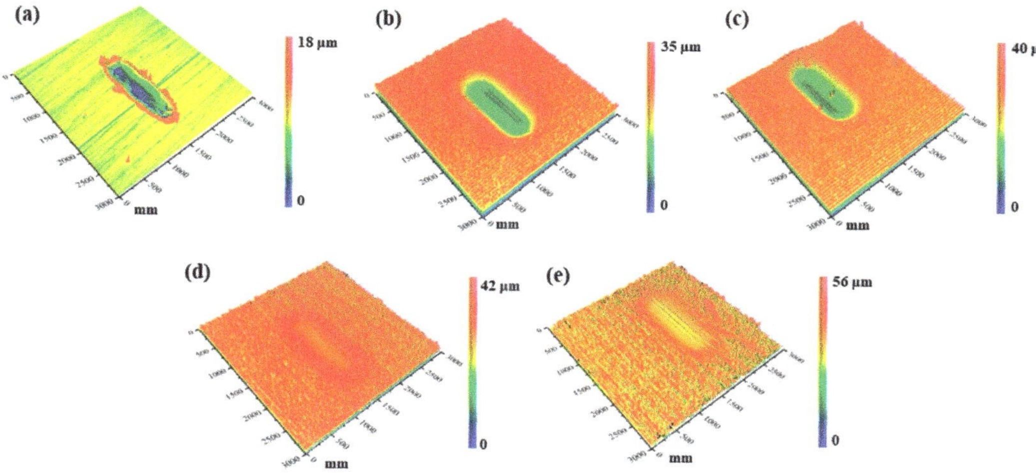

Figure 8. Three-dimensional topography of U and superhydrophobic samples after friction (**a**) U, (**b**) L1, (**c**) L20, (**d**) L50, and (**e**) L100.

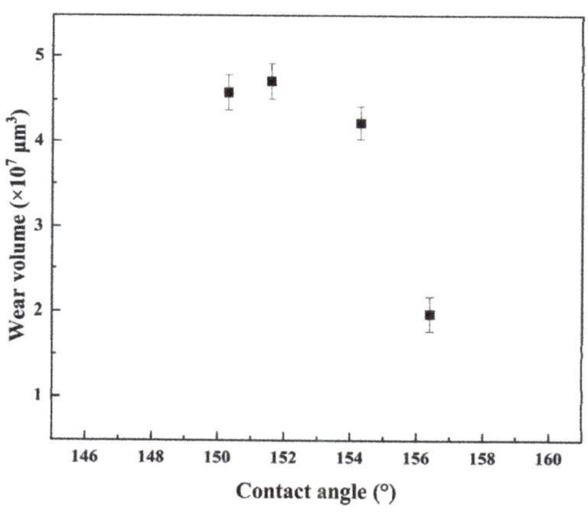

Figure 9. Abrasion volume of U and the superhydrophobic sample surface.

Due to the striped texture, we analyzed the tribological properties of the striped-textured surface in different friction directions. Figure 8 shows that the friction directions of samples L1 and L20 are perpendicular to the stripe texture direction, while the friction directions of L50 and L100 are parallel. It can be seen from Figures 7b and 9 that when the texture is perpendicular to the friction direction, as the contact angle increases, the COF and wear volume increase gradually, but the increase is very small. When the texture is parallel to the friction direction, as the contact angle increases, the COF increases, but the wear volume decreases. The texture perpendicular to the friction direction was more conducive to the preparation of parts that required a small COF and a large wear volume, such as cheap parts that were easy to replace, and which would protect other high-precision and expensive parts. The texture parallel to the friction direction was more conducive to the preparation of situations that required a large COF and small wear volume, such as parts that were not easy to replace or that were expensive.

5. Conclusions

We used a nanosecond pulsed laser to realize the preparation of the superhydrophobic nickel surface and we studied the friction and wear properties of the superhydrophobic nickel surface with different contact angles under dry friction conditions. Compared to the unprocessed surface, the laser-fabricated superhydrophobic surface reduced the friction coefficient of nickel but also increased the wear volume. However, as the contact angle increased, the COF of the superhydrophobic surface increased gradually and the wear volume decreased gradually. The texture perpendicular to the friction direction is more conducive to the preparation of parts requiring a small COF and large wear volume, while the texture parallel to the friction direction is more conducive to the preparation of parts requiring a large COF and small wear volume. This method provides a reference for preparing superhydrophobic surfaces with more suitable friction and wear properties.

Author Contributions: Conceptualization, J.H. and Z.Z.; methodology, J.H.; software, D.G.; validation, Z.Z. and D.G.; formal analysis, Z.N.; investigation, J.H.; resources, L.Z.; data curation, D.G.; writing—original draft preparation, J.H.; writing—review and editing, J.H.; visualization, W.Z.; supervision, L.Z.; project administration, W.Z.; funding acquisition, L.Z. and W.Z. All authors have read and agreed to the published version of the manuscript.

Funding: National Natural Science Foundation of China (51975036).

Institutional Review Board Statement: Not applicable.

Informed Consent Statement: Not applicable.

Data Availability Statement: Not applicable.

Conflicts of Interest: The authors declare no conflict of interest.

References

1. Kung, C.H.; Sow, P.K.; Zahiri, B.; Mérida, W. Assessment and Interpretation of Surface Wettability Based on Sessile Droplet Contact Angle Measurement: Challenges and Opportunities. *Adv. Mater. Interfaces* **2019**, *6*, 1900839. [CrossRef]
2. Sharma, A.; Arora, H.; Grewal, H.S. Self-regenerative superhydrophobic metallic coatings with enhanced durability. *Surf. Coatings Technol.* **2023**, *462*, 129459. [CrossRef]
3. Ding, K.; Wang, C.; Li, S.; Zhang, X.; Lin, N. Large-area cactus-like micro-/nanostructures with anti-reflection and superhydrophobicity fabricated by femtosecond laser and thermal treatment. *Surf. Interfaces* **2022**, *33*, 102292. [CrossRef]
4. Feng, X.; Sun, P.; Tian, G. Recent Developments of Superhydrophobic Surfaces (SHS) for Underwater Drag Reduction Opportunities and Challenges. *Adv. Mater. Interfaces* **2021**, *9*, 2101616. [CrossRef]
5. Lan, X.; Zhang, B.; Wang, J.; Fan, X.; Zhang, J. Hydrothermally structured superhydrophobic surface with superior anti-corrosion, anti-bacterial and anti-icing behaviors. *Colloids Surfaces A Physicochem. Eng. Asp.* **2021**, *624*, 126820. [CrossRef]
6. Liravi, M.; Pakzad, H.; Moosavi, A.; Nouri-Borujerdi, A. A comprehensive review on recent advances in superhy-drophobic surfaces and their applications for drag reduction. *Prog. Org. Coat.* **2020**, *140*, 105537. [CrossRef]
7. Wang, C.X.; Zhang, X.F. A non-particle and fluorine-free superhydrophobic surface based on one-step electrodep-osition of dodecyltrimethoxysilane on mild steel for corrosion protection. *Corros. Sci.* **2020**, *163*, 108284. [CrossRef]
8. Song, Y.; Wang, C.; Dong, X.; Yin, K.; Zhang, F.; Xie, Z.; Chu, D.; Duan, J. Controllable superhydrophobic aluminum surfaces with tunable adhesion fabricated by femtosecond laser. *Opt. Laser Technol.* **2018**, *102*, 25–31. [CrossRef]
9. Liu, Y.; Li, S.; Niu, S.; Cao, X.; Han, Z.; Ren, L. Bio-inspired micro-nano structured surface with structural color and anisotropic wettability on Cu substrate. *Appl. Surf. Sci.* **2016**, *379*, 230–237. [CrossRef]
10. Rajab, F.H.; Liu, Z.; Li, L. Long term superhydrophobic and hybrid superhydrophobic/superhydrophilic surfaces produced by laser surface micro/nano surface structuring. *Appl. Surf. Sci.* **2018**, *466*, 808–821. [CrossRef]
11. Huang, W.; Ordikhani-Seyedlar, R.; Samanta, A.; Shaw, S.; Ding, H. Quantification of superhydrophobic functionali-zation for laser textured metal surfaces. *Colloids Surf. A Physicochem. Eng. Asp.* **2022**, *636*, 128126. [CrossRef]
12. Lu, Y.; Guan, Y.; Li, Y.; Yang, L.; Wang, M.; Wang, Y. Nanosecond laser fabrication of superhydrophobic surface on 316L stainless steel and corrosion protection application. *Colloids Surfaces A Physicochem. Eng. Asp.* **2020**, *604*, 125259. [CrossRef]
13. Huang, J.; Yang, S. Investigation on anisotropic tribological properties of superhydrophobic/superlipophilic lead bronze surface textured by femtosecond laser. *Appl. Surf. Sci.* **2021**, *579*, 152223. [CrossRef]
14. Shimada, H.; Kato, S.; Watanabe, T.; Yamaguchi, M. Direct laser processing of two-scale periodic structures for su-perhydrophobic surfaces using a nanosecond pulsed laser. *Lasers Manuf. Mater. Process.* **2020**, *7*, 496–512. [CrossRef]
15. He, Y.; Wang, L.; Wu, T.; Wu, Z.; Chen, Y.; Yin, K. Facile fabrication of hierarchical textures for substrate-independent and durable superhydrophobic surfaces. *Nanoscale* **2022**, *14*, 9392–9400. [CrossRef]

16. Han, X.; Ren, L.; Ma, Y.; Gong, X.; Wang, H. A mussel-inspired self-repairing superhydrophobic coating with good anti-corrosion and photothermal properties. *Carbon* **2022**, *197*, 27–39. [CrossRef]
17. Li, H.; Luo, Y.; Yu, F.; Zhang, H. In-situ construction of MOFs-based superhydrophobic/superoleophilic coating on filter paper with self-cleaning and antibacterial activity for efficient oil/water separation. *Colloids Surfaces A Physicochem. Eng. Asp.* **2021**, *625*, 126976. [CrossRef]
18. Jiao, Y.; Zhang, T.; Ji, J.; Guo, Y.; Wang, Z.; Tao, T.; Xu, J.; Liu, X.; Liu, K. Functional Microtextured Superhydrophobic Surface with Excellent Anti-Wear Resistance and Friction Reduction Properties. *Langmuir* **2022**, *38*, 13166–13176. [CrossRef] [PubMed]
19. Guo, Y.; Zhang, X.; Wang, X.; Xu, Q.; Geng, T. Wetting and tribological properties of superhydrophobic aluminum surfaces with different water adhesion. *J. Mater. Sci.* **2020**, *55*, 11658–11668. [CrossRef]
20. Huang, J.; Wei, S.; Zhang, L.; Yang, Y.; Yang, S.; Shen, Z. Fabricating the Superhydrophobic Nickel and Improving Its Antifriction Performance by the Laser Surface Texturing. *Materials* **2019**, *12*, 1155. [CrossRef]
21. Kasman, Ş.; Uçar, I.C.; Ozan, S. Investigation of laser surface texturing parameters of biomedical grade Co-Cr-Mo alloy. *Int. J. Adv. Manuf. Technol.* **2023**, *125*, 4271–4291. [CrossRef]
22. Ahangaran, F.; Navarchian, A.H. Recent advances in chemical surface modification of metal oxide nanoparticles with silane coupling agents: A review. *Adv. Colloid Interface Sci.* **2020**, *286*, 102298. [CrossRef] [PubMed]
23. Allahdini, A.; Jafari, R.; Momen, G. Transparent non-fluorinated superhydrophobic coating with enhanced anti-icing performance. *Prog. Org. Coatings* **2022**, *165*, 106758. [CrossRef]
24. Guan, Y.; Cui, X.; Chen, D.; Su, W.; Zhao, Y.; Li, J.; Feng, L.; Li, X.; Jin, G. Realizing excellent tribological properties of BCC/FCC gradient high -entropy alloy coating via an in-situ interface reaction. *Mater. Today Commun.* **2023**, *35*, 106098. [CrossRef]
25. Huang, J.; Guan, Y.; Ramakrishna, S. Tribological behavior of femtosecond laser-textured leaded brass. *Tribol. Int.* **2021**, *162*, 107115. [CrossRef]

hv **photonics**

MDPI

Communication

1.1–1.6 μm Multi-Wavelength Random Raman Fiber Laser

Chunhua Hu [1,2] and Ping Sun [1,*]

1 Shandong Provincial Engineering and Technical Center of Light Manipulations & Shandong Provincial Key Laboratory, School of Physics and Electronics, Shandong Normal University, Jinan 250014, China
2 Electrical Engineering College, Binzhou University, Binzhou 256603, China
* Correspondence: sunping@sdnu.edu.cn

Abstract: Multi-wavelength fiber lasers have attracted great attention due to their application value in many fields. In this work, we demonstrated a seven-wavelength random Raman fiber laser in the range of 1.1–1.6 μm. A piece of 1-km-long high Raman gain optical fiber is utilized as the gain medium. The 1st-order to 7th-order Stokes waves are located, respectively, at 1133 nm, 1194 nm, 1260 nm, 1332 nm, 1414 nm, 1504 nm, and 1606 nm. In the 3-dB bandwidth of optical spectra of 1st-order and 2nd-order Stokes waves, four peaks with an average spacing of 1 nm and 20 peaks with an average spacing of 0.45 nm respectively, are recorded. Pumped by a 1080 nm/12.5 W/220 ns laser, the maximum output power can reach 4.16 W, corresponding to the optical-to-optical conversion efficiency of ~30.7%.

Keywords: multi-wavelength fiber laser; Raman fiber laser; random fiber laser; fiber loop mirror

1. Introduction

Multi-wavelength fiber lasers have attracted extensive attention as a result of their great application value in the fields of multi-species gas monitoring, photoacoustic spectroscopy, wavelength-division-multiplexing (WDM) systems, and fiber-optic sensing, etc. [1–3]. The output characteristics of multi-wavelength fiber lasers rely on the lasing procedures and gain mediums used [4]. The gain mediums will affect working stability and wavelength. The procedures reflect the formation mechanisms of multi-wavelength operations. The types of procedures can be divided into passive elements and active nonlinear processes. Fiber Bragg gratings (FBGs) [5,6], fiber loop mirrors [7], Fabry-Perot [8], and Mach-Zehnder [9] interferometer filters are typical and commonly used passive elements. In addition, the nonlinear processes mainly include Stimulated Brillouin scattering (SBS) [10], Stimulated Raman scattering (SRS) [11], random distribution feedback (RDFB) [4], and four-wave mixing (FWM) [12]. The stable operation of multi-wavelength lasers depends on the cooperation of these nonlinear processes and key elements. Multimode, sampled, and phase-shifted FBGs that can reflect multi-wavelengths can all be applied for multi-wavelength operation. Zhang et al. [13] demonstrated a switchable multi-wavelength thulium-doped fiber laser (TDFL) which can obtain ten-wavelength operation with high stability based on a polarization-maintaining sampled FBG. Fabry-Perot interferometer can generate multi-wavelengths when the cavity length and free spectral ranges satisfy the least common multiple number [14]. Wang et al. [15] reported a multi-wavelength TDFL using a Microfiber-optic Fabry-Perot interferometer with a free spectral range of ~26 nm as the wavelength-selective filter. Sierra-Hernandez et al. [16] presented a multi-wavelength erbium-doped fiber laser based on a Mach-Zehnder interferometer, which is constructed by splicing photonic crystal fiber (PCF) between two segments of a single-mode fiber. Based on the interference between modes, fiber loop mirrors can play the role of fiber filters with low loss to generate multi-wavelength lasing oscillations [4]. Liu et al. [17] obtained 15 output channels within 10-dB bandwidth from 1990 nm to 2007.5 nm by utilizing a fiber Sagnac loop mirror based on polarization-maintaining fiber.

Citation: Hu, C.; Sun, P. 1.1–1.6 μm Multi-Wavelength Random Raman Fiber Laser. *Photonics* **2023**, *10*, 164. https://doi.org/10.3390/photonics10020164

Received: 21 December 2022
Revised: 26 January 2023
Accepted: 29 January 2023
Published: 3 February 2023

Based on the abovementioned passive elements and active nonlinear processes, various kinds of setups like semiconductor optical amplifiers [18,19], rare-earth-doped fiber lasers [20], and Raman fiber lasers (RFLs) [21,22] have demonstrated that they are excellent tools to obtain multi-wavelength operation. Especially, the advent of random Raman fiber lasers (RRFLs) in 2010 [23] suggests the possibility that fiber can be used as a kind of disorder medium to generate multi-wavelength operation under the action of RDFB and Raman amplification via a long distance. RRFLs have greater attraction as they can realize switching of various customized wavelengths by selecting appropriate Raman gain medium, pump source, and other technical parameters. For example, using a homemade tunable Yb-doped fiber laser (1075 ± 25 nm) as the pump source, Balaswamy et al. [24] demonstrated a tuning RRFL with a tuning range from 1450 to 1510 nm, based on a reel of 350-m-length high Raman nonlinearity fiber (effective area 12 μm^2) with special filtering properties. Commonly used single-mode fibers are difficult to use for realizing 2 μm RRFLs limited by the low Raman gain at 2 μm and increased propagation loss. So, highly GeO_2-doped silica fibers (higher Raman gain and lower optical loss at 2 μm) with different lengths are used to explore the performance at 2 μm band [25,26]. In addition, Jin et al. [27] reported a 2.1 μm random fiber laser with 150 m highly GeO_2-doped silica fiber. Moreover, RRFLs have good environmental stability, as the weak RDFB from Rayleigh scattering generates via a long piece of passive fiber. However, the previous research mainly focused on the output of Stokes waves of a certain order based on cascaded Raman shifts. Different techniques and structures are used to improve the purity of output spectra by suppressing the starting vibration of other Stokes waves. Only a few reports focus on the multi-wavelength simultaneous output. Kim et al. [22] reported a tunable multi-wavelength all-fiber Raman source working from 1.12–1.58 μm based on a tunable intracavity high birefringence fiber Sagnac loop filter, but the intensity of each Stokes wave varies greatly. Recently, Adamu et al. [1] built a gas-filled fiber Raman laser which can span from 1.53–2.4 μm. However, the use of a part of 5-m-length hydrogen (H_2)-filled nested anti-resonant fiber, all-polarization maintaining master oscillator power amplifier configuration, and the focusing coupling system makes it complicated and high cost.

In this study, we utilize a piece of 1-km-long Raman optical fiber with a high Raman gain efficiency and low Raman threshold as the gain medium to realize a multi-wavelength operation. This Raman optical fiber can provide enough gain at any wavelength from 1100 to 1700 nm. Based on the high Raman gain fiber, a half-open cavity RRFL is built. Combined with a nanosecond pump laser working at 1080 nm, seven wavelengths of 1.1–1.6 μm can output at the same time under the action of SRS. It should be noted that there are four peaks recorded in the 3-dB bandwidth of the optical spectrum (3.57 nm) of the 1st-order Stokes wave, and 20 peaks in the 3-dB bandwidth of the optical spectrum (8.84 nm) of the 2nd-order Stokes wave.

2. Experimental Setup

In order to realize the possibility of multi-wavelength operation in a wide band of optical spectrum, an RRFL with a half-open cavity structure shown in Figure 1 is built. The Raman gain medium used in our work is a piece of 1-km-long Raman optical fiber (from OFS company) with a high index core and a small effective area which are specially designed to yield a high Raman gain efficiency. The core diameter is about 4 μm and the cladding diameter is about 125 μm. The peak Raman gain efficiency is 2.55 (W·km)$^{-1}$. The effective areas of 1450 nm and 1550 nm are 16.2 μm^2 and 18.94 μm^2. Moreover, the threshold of Raman shift of Raman optical fiber is lower compared with standard single-mode fiber of the same length. The pump laser works at 1080 nm. The maximum pump power and pulse width are 12.5 W and 220 ns. A pump combiner is utilized to inject the pump laser into the random Raman cavity. A fiber loop mirror obtained by fusing two output terminals of an output coupler with an output ratio of 50:50 is connected to the signal part of the pump combiner and used as a total reflection mirror. Under the naturally present random distributed feedback (RDFB) along the 1-km-long Raman optical fiber and the broadband-reflective effect of the fiber loop mirror, Raman optical resonators are formed. The output part is cut with an angle of 8 degrees to avoid unexpected backward reflection. The eventual output wavelengths and corresponding output powers depend entirely on the competition of different wavelengths. To ensure the stable operation of the laser, the whole laser is placed on a water-cooled plate, and the set temperature of the water-cooled plate is 20 °C. The output characteristics are recorded by an optical spectrum analyzer (Anritsu, MS9710C, 0.6–1.75 μm), and an optical power meter (Ophir, Starlite-AC Rohs).

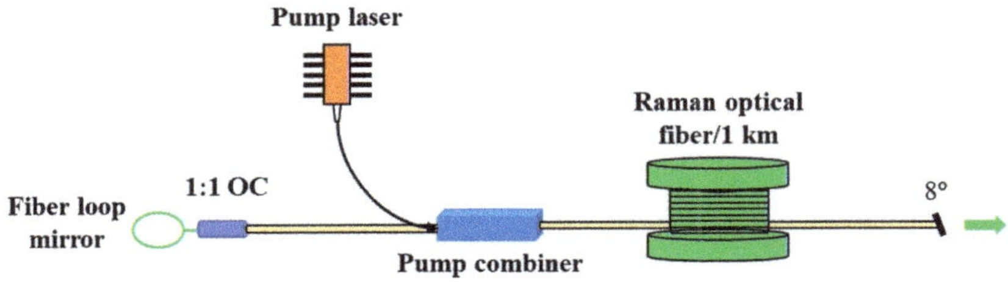

Figure 1. Structure of the multi-wavelength random Raman fiber laser.

3. Experimental Results and Discussion

Based on the setup shown in Figure 1, the multi-wavelength output is realized in our work. Figure 2A exhibits the optical spectrum of the multi-wavelength RRFL. In addition to the fundamental frequency light (pump laser) located at 1080 nm, seven spectral lines from 1st-order to 7th-order Stokes waves corresponding to 1133 nm, 1194 nm, 1260 nm, 1332 nm, 1414 nm, 1504 nm, and 1606 nm can be recorded. The seven Stokes waves can vibrate at the same time since the Raman optical fiber has a low Raman threshold, and the initial pump power (2.6 W) has reached the threshold of any order in the seven Stokes waves. Moreover, the Raman optical fiber can provide relatively uniform Raman gain in the range from 1.1 to 1.7 μm, and the fiber loop mirror does not have the function of selecting specific wavelengths like gratings. So, six Stokes waves can be covered in the intensity range of full width at half maximum (FWHM) of the optical spectrum and have a rather uniform power level. The relation between total output power and pump power is given in Figure 2. The fitting cure has good linearity, and the slope of the fitting curve is ~30.2%. When the pump power is 12.5 W, the output power can reach the maximum value, which is 4.16 W (the residual pump power is about 0.32 W), corresponding to the optical-to-optical conversion efficiency of ~30.7%. Additionally, the output power of different Stokes waves is analyzed, and the experimental results are shown in Figure 3A. As the Raman optical fiber has a large loss for a wavelength over 1.7 μm, higher-order Stokes waves cannot be obtained at the current pump power. The power cannot transfer to higher Stokes waves but only transfer between the seven Stokes waves. Therefore, each Stokes wave will compete with the others, which leads to the intensity ratio of different Stokes waves varying with the change of pump power. On the whole, as the pump power increases, the higher-order Stokes light gradually dominates. First, the fundamental frequency light dominates. Then, the 1st-order Stokes wave begins to dominate, and finally the 6th-order Stokes dominates. The maximum power of 1st-order to 7th-order Stokes waves are ~615 mW, ~418 mW, ~312 mW, ~412 mW, ~642 mW, ~823 mW, and ~613 mW calculated according to the output spectral intensity ratio of different wavelengths.

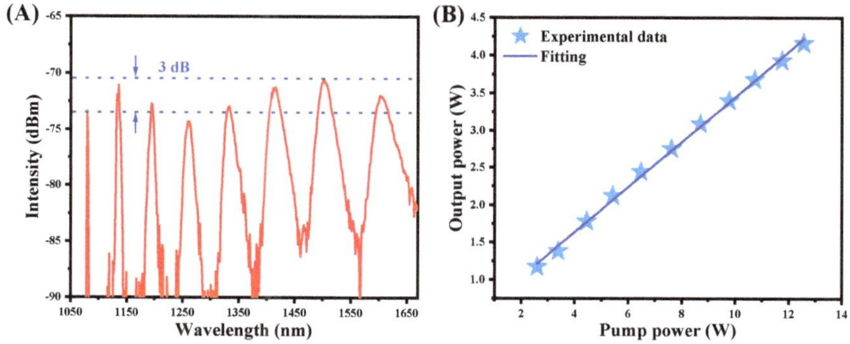

Figure 2. (**A**) Optical spectrum of the multi-wavelength RRFL; (**B**) Total output power of the multi-wavelength RRFL.

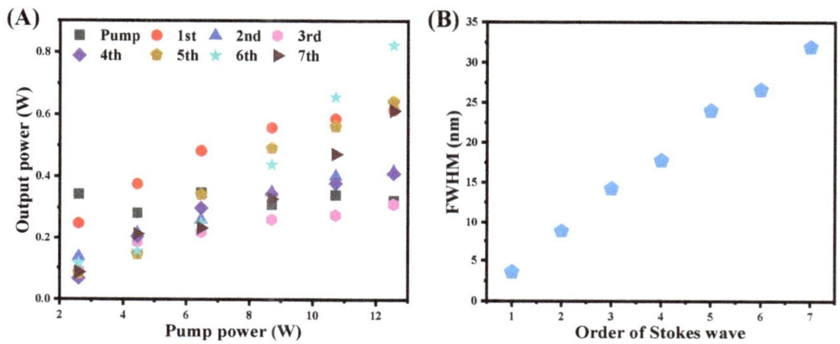

Figure 3. (**A**) Output power of different Stokes waves; (**B**) FWHM of different Stokes waves.

The optical spectra of each Stokes wave are tested and analyzed. The FWHM of the optical spectrum increases with increases in Stokes waves' order. As shown in Figure 3B, the FWHM of the 7th-order Stokes wave can reach 31.9 nm, and that of the 1st-order Stokes wave is only 3.57 nm. In particular, the multi-wavelength phenomenon is not only obtained in a wide range from 1.1–1.6 μm but also observed in the optical spectra of 1st-order and 2nd-order Stokes waves. And Figure 4A,B, respectively, exhibit their fine optical spectra under a resolution of 0.05 nm. For the 1st-order Stokes wave, four wavelength peaks are recorded in the 3-dB bandwidth of the optical spectrum, which is 3.57 nm, and the average wavelength spacing is 1 nm. For the 2nd-order Stokes wave, there are 20 peaks in its 3-dB bandwidth of the optical spectrum, which is 8.84 nm, and the average wavelength spacing decreases to 0.45 nm. However, this phenomenon is not observed in other Stokes waves, which is mainly due to the spectral broadening of high-order Stokes waves making these subtle modulations difficult to distinguish as the peaks with small spacing connected together in the process of Raman frequency shift. As we know, the formation mechanism of 1st-order to 7th-order Stokes waves is stimulated Raman scattering. However, the peaks located on the 1st-order and 2nd-order Stokes waves are mainly caused by the filtering effect of fiber loop mirror combined with the random distributed feedback from the 1-km-long Raman optical fiber and FWM [28,29], which is different from the multi-wavelength operation of the seven Stokes waves.

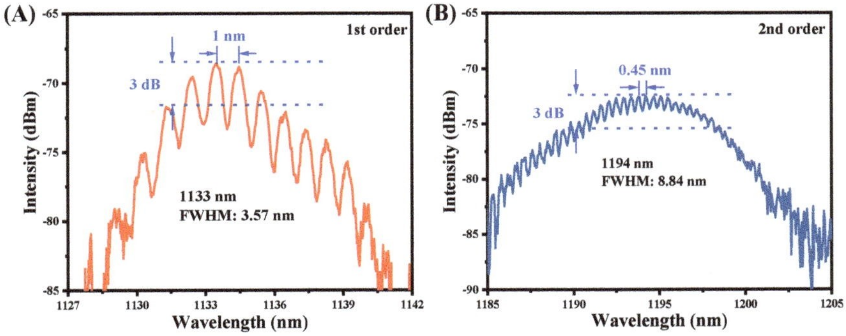

Figure 4. (**A**) Optical spectrum of the 1st-order Stokes wave; (**B**) Optical spectrum of the 2nd-order Stokes wave.

4. Conclusions

In conclusion, a seven-wavelength random Raman fiber laser in the range of 1.1–1.6 μm is realized in this work. The 1st-order to 7th-order Stokes waves are located at 1133 nm, 1194 nm, 1260 nm, 1332 nm, 1414 nm, 1504 nm, and 1606 nm, respectively, corresponding to the maximum output power of ~615 mW, ~418 mW, ~312 mW, ~412 mW, ~642 mW, ~823 mW, and ~613 mW. In the 3 dB bandwidth of optical spectra of 1st-order and 2nd-order Stokes waves, four peaks with an average spacing of 1 nm and 20 peaks with an average spacing of 0.45 nm are recorded, respectively. This work is expected to be applied in the fields of multi-species gas monitoring, WDM, and fiber sensors through subsequent improvement and testing.

Author Contributions: Conceptualization, C.H. and P.S.; methodology, C.H. and P.S.; validation, C.H.; formal analysis, C.H.; investigation, C.H.; resources, P.S.; writing—original draft preparation, C.H.; writing—review and editing, P.S.; supervision, P.S.; project administration, P.S.; funding acquisition, P.S. All authors have read and agreed to the published version of the manuscript.

Funding: This research was funded by National Natural Science Foundation of China (61975099).

Institutional Review Board Statement: Not applicable.

Informed Consent Statement: Not applicable.

Data Availability Statement: The data presented in this study are available upon reasonable request.

Conflicts of Interest: The authors declare no conflict of interest.

References

1. Adamu, A.I.; Wang, Y.; Habib, M.S.; Dasa, M.K.; Antonio-Lopez, J.E.; Amezcua-Correa, R.; Bang, O.; Markos, C. Multi-wavelength high-energy gas-filled fiber Raman laser spanning from 1.53 μm to 2.4 μm. *Opt. Lett.* **2021**, *46*, 452–455. [CrossRef]
2. Sugavanam, S.; Yan, Z.; Kamynin, V.; Kurkov, A.S.; Zhang, L.; Churkin, D.V. Multiwavelength generation in a random distributed feedback fiber laser using an all fiber Lyot filter. *Opt. Express* **2014**, *22*, 2839–2844. [CrossRef] [PubMed]
3. Wang, Z.N.; Wu, H.; Fan, M.Q.; Li, Y.; Gong, Y.; Rao, Y.J. Broadband flat-amplitude multiwavelength Brillouin-Raman fiber laser with spectral reshaping by Rayleigh scattering. *Opt. Express* **2013**, *21*, 29358–29363. [CrossRef] [PubMed]
4. Du, X.Y.; Zhang, H.W.; Wang, X.; Wang, X.L.; Zhou, P.; Liu, Z.J. Multiwavelength Raman fiber laser based on polarization maintaining fiber loop mirror and random distributed feedback. *Laser Phys. Lett.* **2015**, *12*, 45106. [CrossRef]
5. Fernandez-Vallejo, M.; Díaz, S.; Perez-Herrera, R.A.; Passaro, D.; Selleri, S.; Quintela, M.A.; López Higuera, J.M.; Lopez-Amo, M. Resilient long-distance sensor system using a multiwavelength Raman laser. *Meas. Sci. Technol.* **2010**, *21*, 94017. [CrossRef]
6. Han, Y.G.; Tran, T.V.; Kim, S.H.; Lee, S.B. Multiwavelength Raman-fiber-laser-based long-distance remote sensor for simultaneous measurement of strain and temperature. *Opt. Lett.* **2005**, *30*, 1282–1284. [CrossRef]
7. Luo, Z.Q.; Cai, Z.P.; Huang, J.F.; Ye, C.C.; Huang, C.H.; Xu, H.Y.; Zhong, W.D. Stable and spacing-adjustable multiwavelength Raman fiber laser based on mixed-cascaded phosphosilicate fiber Raman linear cavity. *Opt. Lett.* **2008**, *33*, 1602–1604. [CrossRef] [PubMed]
8. Pan, S.L.; Lou, C.Y.; Gao, Y.Z. Multiwavelength erbium-doped fiber laser based on inhomogeneous loss mechanism by use of a highly nonlinear fiber and a Fabry-Perot filter. *Opt. Express* **2006**, *14*, 1113–1118. [CrossRef] [PubMed]

9. Wang, F.; Zhang, X.L.; Yu, Y.; Huang, X. 82-channel multi-wavelength comb generation in a SOA fiber ring laser. *Opt. Laser Technol.* **2010**, *42*, 285–288. [CrossRef]
10. Yuan, Y.J.; Yao, Y.; Yi, M.; Guo, B.; Tian, J.J. Multiwavelength fiber laser employing a nonlinear Brillouin optical loop mirror: Experimental and numerical studies. *Opt. Express* **2014**, *22*, 15352–15363. [CrossRef]
11. Shirazi, M.R.; Mohamed Taib, J.; Dimyati, K.; Harun, S.W.; Ahmad, H. Multi-wavelength Brillouin–Raman fiber laser generation assisted by multiple four-wave mixing processes in a ring cavity. *Laser Phys.* **2013**, *23*, 75108. [CrossRef]
12. Ahmad, H.; Parvizi, R.; Dimyati, K.; Tamjis, M.R.; Harun, S.W. FWM-based multi-wavelength erbium-doped fiber laser using Bi-EDF. *Laser Phys.* **2010**, *20*, 1414–1417. [CrossRef]
13. Zhang, L.N.; Yan, F.P.; Feng, T.; Guo, Y.; Qin, Q.; Zhou, H.; Suo, Y.P. Switchable Multi-Wavelength Thulium-Doped Fiber Laser Employing a Polarization-Maintaining Sampled Fiber Bragg Grating. *IEEE Access* **2019**, *7*, 155437–155445. [CrossRef]
14. Yeh, C.H.; Chow, C.W.; Wu, Y.F.; Shih, F.Y.; Wang, C.H.; Chi, S. Multiwavelength erbium-doped fiber ring laser employing Fabry–Perot etalon inside cavity operating in room temperature. *Opt. Fiber Technol.* **2009**, *15*, 344–347. [CrossRef]
15. Wang, M.; Huang, Y.J.; Yu, L.; Song, Z.P.; Liang, D.Z.; Ruan, S.C. Multiwavelength Thulium-Doped Fiber Laser Using a Micro Fiber-Optic Fabry–Perot Interferometer. *IEEE Photon. J.* **2018**, *10*, 1–8. [CrossRef]
16. Sierra-Hernandez, J.M.; Rojas-Laguna, R.; Vargas-Rodriguez, E.; Estudillo-Ayala, J.M.; Mata-Chavez, R.I.; Jauregui-Vazquez, D.; Hernandez-Garcia, J.C.; Andrade-Lucio, J.A.; Gutierrez-Gutierrez, J.C. A tunable multi-wavelength laser based on a Mach–Zehnder interferometer with photonic crystal fiber. *Laser Phys.* **2013**, *23*, 55105. [CrossRef]
17. Liu, S.; Yan, F.P.; Wu, B.L.; Tan, S.Y.; Peng, W.J.; Feng, T.; Liang, X.; Li, Q. A multiwavelength thulium-doped silica fiber laser incorporating a highly nonlinear fiber. *J. Opt.* **2014**, *16*, 55201. [CrossRef]
18. Liu, D.; Ngo, N.Q.; Liu, H.; Liu, D. Stable multiwavelength fiber ring laser with equalized power spectrum based on a semiconductor optical amplifier. *Opt. Commun.* **2009**, *282*, 1598–1601. [CrossRef]
19. Pleros, N.; Bintjas, C.; Kalyvas, M.; Theophilopoulos, G.; Yiannopoulos, K.; Sygletos, S.; Avramopoulos, H. Multiwavelength and power equalized SOA laser sources. *IEEE Photon. Technol. Lett.* **2002**, *14*, 693–695. [CrossRef]
20. Liang, D.M.; Xu, X.F.; Li, Y.; Pei, J.H.; Jiang, Y.; Kang, Z.H.; Gao, J.Y. Multiwavelength fiber laser based on a high-birefringence fiber loop mirror. *Laser Phys. Lett.* **2007**, *4*, 57–60. [CrossRef]
21. Frazão, O.; Correia, C.; Santos, J.L.; Baptista, J.M. Raman fibre Bragg-grating laser sensor with cooperative Rayleigh scattering for strain–temperature measurement. *Meas. Sci. Technol.* **2009**, *20*, 45203. [CrossRef]
22. Kim, C.S.; Sova, R.M.; Kang, J.U. Tunable multi-wavelength all-fiber Raman source using fiber Sagnac loop filter. *Opt. Commun.* **2003**, *218*, 291–295. [CrossRef]
23. Turitsyn, S.K.; Babin, S.A.; El-Taher, A.E.; Harper, P.; Churkin, D.V.; Kablukov, S.I.; Ania-Castañón, J.D.; Karalekas, V.; Podivilov, E.V. Random distributed feedback fibre laser. *Nat. Photonics* **2010**, *4*, 231–235. [CrossRef]
24. Balaswamy, V.; Arun, S.; Aparanji, S.; Choudhury, V.; Supradeepa, V.R. High-power, fixed, and tunable wavelength, grating-free cascaded Raman fiber lasers. *Opt. Lett.* **2018**, *43*, 1574–1577. [CrossRef] [PubMed]
25. Jiang, H.; Zhang, L.; Feng, Y. Silica-based fiber Raman laser at >2.4 μm. *Opt. Lett.* **2015**, *40*, 3249–3252. [CrossRef]
26. Liu, J.; Tan, F.; Shi, H.; Wang, P. High-power operation of silica-based Raman fiber amplifier at 2147 nm. *Opt. Express* **2014**, *22*, 28383–28389. [CrossRef]
27. Jin, X.; Lou, Z.; Zhang, H.; Xu, J.; Zhou, P.; Liu, Z. Random distributed feedback fiber laser at 2.1 μm. *Opt. Lett.* **2016**, *41*, 4923–4926. [CrossRef]
28. El-Taher, A.E.; Harper, P.; Babin, S.A.; Churkin, D.V.; Podivilov, E.V.; Ania-Castanon, J.D.; Turitsyn, S.K. Effect of Rayleigh-scattering distributed feedback on multiwavelength Raman fiber laser generation. *Opt. Lett.* **2011**, *36*, 130–132. [CrossRef]
29. Pinto, A.M.R.; Frazão, O.; Santos, J.L.; Lopez-Amo, M. Multiwavelength fiber laser based on a photonic crystal fiber loop mirror with cooperative Rayleigh scattering. *Appl. Phys. B* **2010**, *99*, 391–395. [CrossRef]

photonics

MDPI

Article

An Ytterbium-Doped Narrow-Bandwidth Randomly Distributed Feedback Laser Emitting at a Wavelength of 976 nm

Danila A. Davydov [1], Andrey A. Rybaltovsky [1], Svetlana S. Aleshkina [1], Vladimir V. Velmiskin [1], Mikhail E. Likhachev [1,*], Sergei M. Popov [2], Dmitry V. Ryakhovskiy [2], Yuriy K. Chamorovskiy [2], Andrey A. Umnikov [3] and Denis S. Lipatov [3]

1 Prokhorov General Physics Institute of the Russian Academy of Sciences, Dianov Fiber Optics Research Center, 119333 Moscow, Russia; danila-davydov.2015@mail.ru (D.A.D.); rybaltovsky@yandex.ru (A.A.R.); sv_alesh@fo.gpi.ru (S.S.A.); vvv@fo.gpi.ru (V.V.V.)

2 Kotelnikov Institute of Radioengineering and Electronics (Fryazino Branch), Russian Academy of Sciences, 1 Akad. Vvedenskogo Pl., Fryazino, 141190 Moscow Region, Russia; sergei@popov.eu.org (S.M.P.); yurichamor@gmail.com (Y.K.C.)

3 Devyatykh Institute of Chemistry of High-Purity Substances of the Russian Academy of Sciences, 49 Tropinina St., 603951 Nizhny Novgorod, Russia

* Correspondence: likhachev@fo.gpi.ru

Abstract: All-fiber, polarization maintaining, narrow-bandwidth, Yb-doped fiber lasers with randomly distributed feedback operated near 976 nm were realized for the first time. It was shown that the laser operated in a single, longitudinal mode regime during intervals of a few seconds. At other times, the laser generated a few longitudinal modes, but its bandwidth was always below the resolution of the optical spectrum analyzer (0.02 nm). The linewidth of each single longitudinal mode of the laser was estimated to be below 20 kHz. The reasons for this observed laser behavior were discussed and methods for achieving stable, continuous wave operation in the single-longitudinal-mode regime were proposed.

Keywords: fiber laser; random distributed feedback; ytterbium-doped fiber; fiber Bragg grating; single-frequency laser

Citation: Davydov, D.A.; Rybaltovsky, A.A.; Aleshkina, S.S.; Velmiskin, V.V.; Likhachev, M.E.; Popov, S.M.; Ryakhovskiy, D.V.; Chamorovskiy, Y.K.; Umnikov, A.A.; Lipatov, D.S. An Ytterbium-Doped Narrow-Bandwidth Randomly Distributed Feedback Laser Emitting at a Wavelength of 976 nm. *Photonics* **2023**, *10*, 951. https://doi.org/10.3390/photonics10080951

Received: 23 June 2023
Revised: 28 July 2023
Accepted: 16 August 2023
Published: 19 August 2023

1. Introduction

Narrow-bandwidth, ytterbium-doped fiber lasers, emitting at a wavelength near 976 nm, are attracting increasing attention due to being promising for a number of applications. Thus, narrow-bandwidth emission at the given wavelength can be required for the development of highly stable, single-frequency, ytterbium- or erbium-doped fiber lasers [1]. A wider range of applications [2–4] opens up new perspectives by doubling the radiation of such lasers (488–490 nm, blue-green radiation). In this case, the output wavelength of the laser is the same as the emission wavelength of an Ar laser, which is relatively inefficient, expensive, has a short lifetime, contains many bulk elements, and requires alignment by highly qualified personnel. Therefore, the frequency doubling of lasing at 976 nm can also be considered as an alternative to the argon laser [5–7]. Further doubling of the lasing frequency makes it possible to obtain UV radiation in the wavelength region of 244 nm [8], which is a promising source for recording fiber Bragg gratings (FBG) and lithography [9,10]. It is worth noting that existing techniques for the doubling of the frequency of a continuous wave (CW) signal require the use of narrow-bandwidth radiation (this is the case of periodically polished crystals [5–7]), or even single-frequency radiation (when doubling with the help of high-Q Fabry–Perot resonators) [8,11].

As a rule, single-frequency lasers are in high demand, the output power of which is units and even tens of watts. In general, such laser systems are built according to the MOPA (master oscillator power amplifier) principle. In this case, the manufacture of both a sufficiently narrow-bandwidth master laser and a highly efficient amplifier is a great challenge. The main problem sin the implementation of an Yb-doped fiber amplifier emitting at 976 nm are a sharp increase in amplified spontaneous emission (ASE) and the appearance of competing generation in the spectral region, with a maximum at 1030 nm with increasing pump power. To suppress radiation in the wavelength region above 1 μm, as a rule, at the stage of the implementation of the laser scheme, the length of the active fiber is reduced, which, in turn, leads to a larger fraction of the unabsorbed pump power and a relatively low total efficiency of converting the pump radiation into a signal. For this reason, specially designed active fibers are used to achieve an acceptably high pump-to-signal conversion efficiency. In particular, this goal can be achieved by introducing excessive optical losses into the spectral region of 1030 nm [12,13], or by creating fibers with an increased ratio of the area of the core to the cross-section area of the first reflective cladding [14–18].

The key element of a single-frequency laser is the master oscillator, which determines the main output characteristics (linewidth, reliability, and stability, etc.). As a master laser, it is possible to use a conventional semiconductor distributed feedback (DFB) diode. However, standard diodes have a linewidth of several MHz, and achieving a narrower linewidth requires the use of rather complex and expensive designs of such lasers. A more elegant solution is to create all-fiber, single-frequency lasers, due to their potentially higher reliability and resistance to external electromagnetic influences. When using relatively simple designs, such lasers can exhibit a narrower emission line by an order of magnitude [19]. Thus, relatively recently, the sub-kHz linewidth was demonstrated in a rather simple and reliable design of a single-frequency fiber laser [20]. The main problem of creating single-frequency fiber lasers is that, in their most common designs, namely lasers with distributed Bragg reflection (DBR) and DFB fiber lasers, the length of the active fiber is extremely short (~1 cm). This is due to the fact that the spectral distance between the longitudinal modes of the resonator must exceed the width of the reflection spectrum of the Bragg gratings. In most cases, this length of an active fiber is not enough to absorb a significant fraction of the pump, and as a result, the efficiency of such lasers is quite low. Only recent work has made it possible to improve the efficiency of such lasers by using heavily doped fibers [21,22].

The problem of creating all-fiber single-frequency lasers in the spectral region near 976 nm is that single-mode diodes emitting at a wavelength of ~915 nm are used as a pump source. The absorption of ytterbium at this wavelength is four times less than that at a wavelength of 976 nm (which is used to pump ytterbium lasers in the region of 1030–1064 and for Er-Yb single-frequency lasers), which makes it much more difficult to overcome the lasing threshold at a wavelength of 976 nm. At present, only the use of ytterbium-doped phosphate glasses has made it possible to obtain single-frequency generation in all-fiber lasers at a wavelength of 976 nm [23,24]. At the same time, the use of such fibers is by no means always acceptable for practical purposes, due to the low thermal and mechanical stability of phosphate glasses. To the best of our knowledge, single-frequency, all-fiber lasers at a wavelength of 976 nm based on silica optical fibers have not yet been realized.

The key to solving this problem may be the creation of a laser with randomly distributed feedback [25–29]. In such lasers, feedback is provided by Rayleigh reflection from an extended section of the fiber, or from an array of low-reflecting FBGs written along the length of the fiber. In the latter case, lasers are capable of generating an extremely narrow spectral line of <6 kHz in the CW mode [30,31]. The advantage of such lasers is the possibility of using the active medium itself for recording gratings, which makes it possible to significantly increase the length of the active fiber inside the resonator, thereby dramatically increasing the efficiency of the laser.

The purpose of this work was to study the particularities of creating a polarized, all-fiber laser with distributed random feedback at a wavelength of about 976 nm and to study the properties of such lasers. In particular, a random-distributed feedback laser was built using polarization maintaining elements to achieve a low-lasing threshold with only one polarization. A detailed study of the regime of laser generation and its dependence on the laser cavity length was performed. The possibility of achieving a single-frequency, single-polarization operation regime near 976 nm with a very narrow bandwidth (less than 20 kHz) for a relatively long time (few seconds) was demonstrated for the first time.

2. Materials and Methods

To develop a laser with random feedback, we used the laser design implemented in our previous work [32]. The main difference of the current work was that the circuit was implemented using polarization-maintaining optical fibers, which is necessary to achieving a single-polarization regime. For these purposes, we used the same initial preform of an ytterbium-doped fiber as in [32]. The preform was made using the modified chemical vapor deposition technique and its core was made of phosphogermanosilicate glass, which has a high photosensitivity. The concentration of ytterbium oxide corresponded to the core absorption at a level of 4 dB/m @ 915 nm. The fiber was made in the standard "PANDA" configuration—two holes were drilled into the silica preform symmetrically with respect to the core, and before drawing the fiber, two borosilicate rods were inserted into these holes. The presence of boron stress rods in the cladding created birefringence in the fiber core, which, in turn, resulted in the appearance ability of the fiber to maintain polarization. Similar to [32], weakly reflecting FBGs were recorded in the fiber directly during drawing by the use of a UV excimer laser. The estimated size of each of the recorded gratings was about 1 cm (corresponding to the width of the phase mask used to record the gratings). The distance between the recorded gratings was also about 1 cm (that is, the gratings were recorded practically without gaps between them). The estimated reflection coefficient of a single grating was about 0.05%. Due to the presence of birefringence, the reflection wavelengths of the gratings for the fast and slow axes differed and amounted to 976.25 nm and 976.5 nm, respectively (see Figure 1).

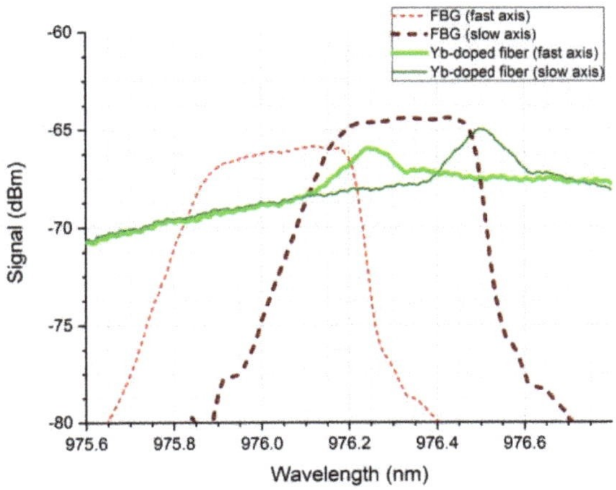

Figure 1. Reflected signal from the FBGs written in the Yb-doped fiber and from separate FBG written in polarization-maintaining fiber. Measurements were made with polarized wideband source to separately characterize fast and slow polarization in both cases. The bold lines correspond to those FBGs, which form the resonator in the realized laser operated at 976.2 nm.

The scheme of the realized laser with random feedback is depicted in Figure 2. The polarization-maintaining, ytterbium-doped fiber with a written array of FBGs (YDF-FBG-PM) was pumped by a semiconductor laser diode (pump diode), with a wavelength of 910 nm and an average power of about 200 mW, through a wavelength division multiplexor (WDM), separating the wavelengths of 910 nm and 976 nm. The 976 nm WDM port was used as the laser output and was spliced to an Isolator, which allowed only one of the polarizations to pass through.

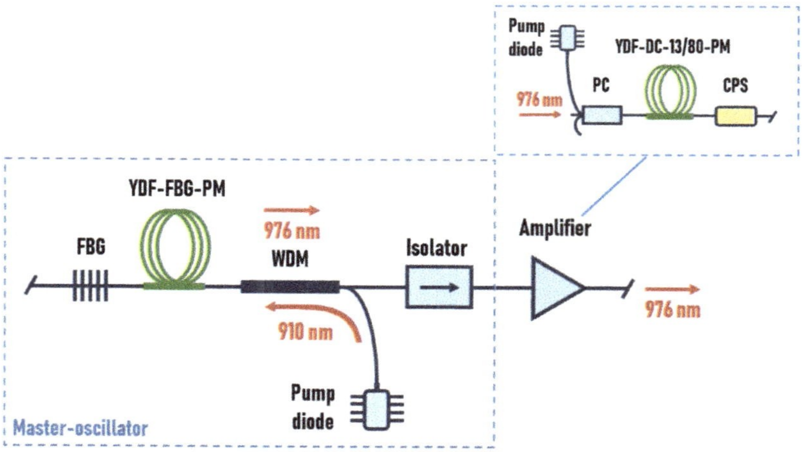

Figure 2. Schematic of a fiber laser with random feedback and a weak-signal amplifier.

A highly reflective FBG written in the polarization-maintaining fiber was spliced to the free end of the ytterbium fiber. The wavelengths corresponding to the reflection maximum of this FBG also differed for "fast" and "slow" polarizations (see Figure 1) and amounted to 975.9–976.2 nm and 976.2–976.4 nm, respectively. During the splicing, the "fast" axis of the YDF-FBG-PM was aligned with the "slow" axis of the YDF-FBG-PM. As a consequence, this made it possible to implement a high-quality resonator for the "fast" polarization of the YDF-FBG-PM—the Bragg grating for this polarization provided high reflection at wavelengths of 976.2–976.4 nm, and the array of Bragg gratings recorded in the YDF-FBG-PM provided random feedback at wavelengths near 976.25 nm. On the contrary, the "slow" polarization of the YDF-FBG-PM (with maximum reflection at 976.5 nm) was aligned to the fast polarization of the FBG (with reflection at 975.9–976.2 nm). Due to mismatch of the reflection spectrum, the quality factor of the resonator for this polarization was low, and lasing in this polarization was suppressed. As a result, conditions were created that ensured the generation of radiation in only one polarization—without the above-mentioned efforts, lasing could occur in both polarizations [33].

During the first experiment, only the master oscillator (shown in Figure 2 by a dashed rectangle) was characterized. In this case, all the free fiber ends were cleaved at an angle of about 7 degrees in order to avoid unwanted Fresnel back reflections from the corresponding ends. In the second experiment after the isolator, a weak-signal amplifier similar to that developed in [17] was placed.

3. Results

First, we studied the effect of the length of the ytterbium-doped fiber with a recorded array of Bragg gratings (YDF-FBG-PM) on the output characteristics of the developed laser. For this purpose, a 3 m long YDF-FBG-PM was gradually shortened, and for each length of the YDF-FBG-PM, the output signal spectra, oscillograms, polarization extinction ratio (PER), power, and stability for several minutes were measured. To measure the PER of the laser, two measurements of the laser radiation power at the output of the isolator were carried out—one when the slow axis of the isolator was aligned with the "fast" polarization of the YDF-FBG-PM at the output of the WDM, and the second when it was aligned with the "slow" polarization. As the fast axis of the isolator was blocked, it acted as a polarizer. This made it possible to determine the signal power propagating in the "fast" and "slow" polarization, respectively, and to determine the PER.

The main results measured with different lengths of the YDF-FBG-PM fiber are shown in Figure 3. It can be seen that the maximum output power increased with an increase in the length of the YDF-FBG-PM, which was associated with an increase in the fraction of the absorbed pump in this fiber. At the same time, with an increase in the length of the YDF-FBG-PM over 1 m, the PER deteriorated significantly. We suggest that this was due to the fact that the intrinsic reflectivity of the array of Bragg gratings recorded in the YDF-FBG-PM became sufficient for achieving efficient generation in the unwanted polarization. This assumption was partly confirmed by the measurement of the output radiation spectrum—in the long-wavelength region (976.5 nm), an additional peak began to appear, approximately coinciding with the reflection peak of the array of Bragg gratings for unwanted polarization (see Figure 4a). Moreover, the output spectra became unstable and varied with time—in Figure 4a, three spectra measured with time intervals of a few seconds are shown (signal #1, signal #2, and signal #3). It is also necessary to note that an increase in the ASE in the wavelength region of 976 nm was observed at the maximum length of the active fiber (3 m). Reducing the length of the ytterbium-doped fiber to 0.7 m led to a sharp drop in the output power to a level of less than 1 mW and an increase in the instability of the output radiation by an order of magnitude (variations in the output power over 10 min increased from 2% to 25%). Changes in the lasing wavelength within 0.1 nm were also observed with time. Apparently, with such a length of the active fiber, the available pump power was not enough to ensure stable laser generation.

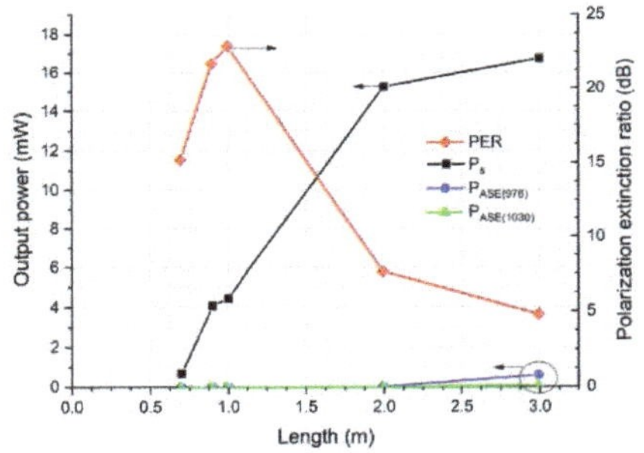

Figure 3. Dependence of the output power of the signal (P_s) at 976.2 nm, the ASE level in the wavelength region of 976 nm ($P_{ASE(976)}$) and at the wavelength of 1030 nm ($P_{ASE(1030)}$), and the PER value against the length of the ytterbium-doped fiber with the recorded array of Bragg gratings.

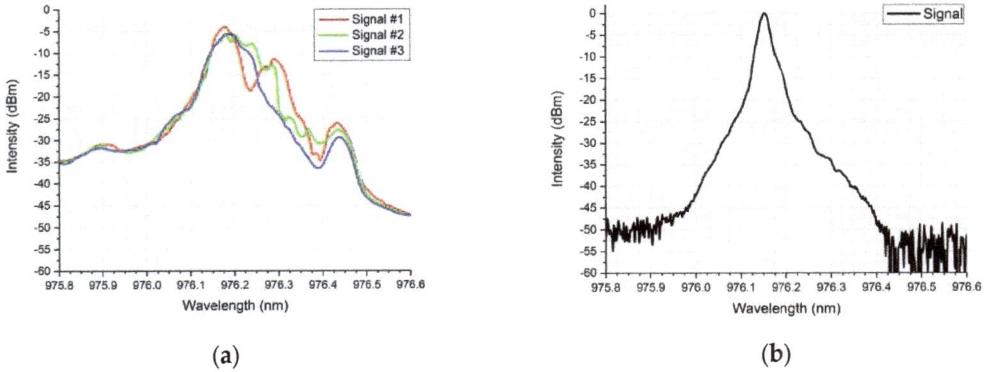

Figure 4. Spectra of output signal for the Yb-doped fiber length of 3 m (**a**) and for the fiber length of 0.9 m (**b**) measured with optical spectrum analyzer ANDO AQ6317B.

As a result, the optimal length of the fiber was found to be in the range of 0.9–1 m—in this case, an acceptable high output power (about 4 mW) and maximum PER level were obtained. Also in this case, the most stable lasing spectrum was observed (see Figure 4b)—within the measurement error of the spectrum analyzer ANDO AQ6317B, the position of the line did not change, and the measured linewidth was 0.02 nm (by 3 dB level), which corresponded to the maximum resolution of the spectrum analyzer at the given wavelength.

In the framework of this research, we did not have the technical ability to directly measure the linewidth of the resulting laser with random feedback. However, an analysis of the oscillograms of radiation at the output of the laser was quite informative. Indeed, if we imagine the developed laser as a laser with a Fabry–Perot resonator, where one of the mirrors is a highly reflective Bragg grating, and the second is an array of weakly reflecting gratings (in the first approximation, they can be represented as one grating with a complex reflection spectrum located in the middle of the active fiber), then the spectral distance between the modes of this resonator can be estimated in orders of magnitude. Equations for the intrinsic frequency of the so-called longitudinal mode (ν_q):

$$\nu_q = \frac{c}{n \cdot \lambda} = q \cdot \frac{c}{2 \cdot n \cdot L}, \ \Delta\nu_q = \frac{c}{2 \cdot n \cdot L}, \tag{1}$$

where c—the light speed velocity, n—the effective refractive index of the fundamental mode in the YDF-FBG-PM, λ—the wavelength of the longitudinal mode with index q, L—the resonator length, and $\Delta\nu_q$—the frequency difference between two neighboring longitudinal modes. In accordance with Equation (1), the distance between the modes can be estimated as 200 MHz for a resonator with an active fiber segment of 0.9 m (the L was taken to be ~0.45 m). The above analysis of the properties of the created resonator was based on a very rough simplification. An exact calculation of the eigenmodes of the created resonator is rather complicated, as it requires the consideration of the interference between all the FBGs (~100 items), which form the laser resonator. Nevertheless, as will be seen below, this approach made it possible to estimate, with an acceptable accuracy, the distance between the resonator eigenmodes.

The gain band of the active medium in our case had a width that was orders of magnitude greater than the distance between the longitudinal modes. However, the reflection spectrum of an array of Bragg gratings due to internal interference should have narrow reflection maxima randomly located along the spectrum that differ in intensity. Exactly within these narrow lines, lasing will develop, provided that the resonator mode falls within these maxima. At the same time, it seems very likely that there will be several modes within the maximum reflection limit of the array of Bragg gratings, which will lead to the generation of several modes. In this case, it will be possible to observe interference between the generated modes—sinusoidal modulations of the output power at half of the frequency spacing between the modes (~100 MHz, 200 MHz, 200 MHz, and so on) will appear on the oscillogram. Given the estimated distance between the modes, the presence of output power modulation when generating several modes can be observed using an oscilloscope with a bandwidth greater than the frequency difference between the modes. We used a Tektronix TDS 3054C oscilloscope with a bandwidth of 500 MHz, which made it possible to carry out these measurements.

The results of measuring the oscillograms in the developed laser with an active fiber length of 0.9 m are shown in Figure 5. It was found that the fabricated laser's significant time (about few seconds) operated in a quasi-continuous mode—(Regime 1). Periodically, this regime was switched to Regime 2 when modulation was observed with a frequency of ~100 MHz, corresponding to the beat between neighboring longitudinal modes of the resonator and even higher (~180 MHz and 280 MHz, etc.).

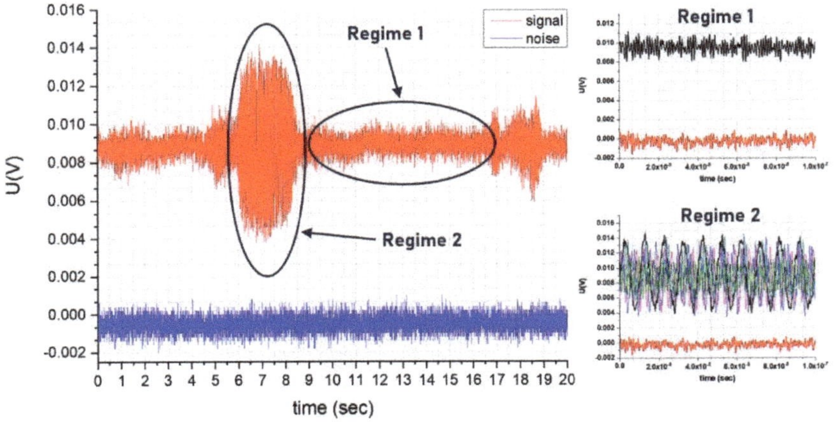

Figure 5. Signals obtained by means of an oscilloscope with the YDF-FBG-PM length of 0.9 m. Curves with different colors for Regime 2 was measured at different times to show different forms of signal, observed in the Regime 2.

The presence of a single longitudinal mode operation regime was also confirmed by the analysis of the output signal using a real-time radio frequency spectrum analyzer (KEYSIGHT EXA Signal Analyzer) with an operating broadband frequency range from 10 Hz to 3.6 GHz. Indeed, in Regime 1, there was no beating between the modes (see Figure 6a). On the contrary, very different radiofrequency spectra were observed in Regime 2—from time to time, oscillation at different frequencies appeared (see Figure 6b). In all cases, the smallest oscillation frequency was near 100 MHz (corresponding to the interference between two neighboring longitudinal modes), the next oscillation was observed at ~190 MHz (double minimum distance between longitudinal modes), and so on (~290 MHz, 300 MHz, 380 MHz, 450 MHz, 550 MHz, and 630 MHz). The exact position of each line changed over time—for example, the position of the first oscillation ranged from 98 MHz to 104 MHz. An example of such variation in a smaller spectral region is shown in Figure 6c. Such variations are a feature of lasers with random distributed feedback—the minimum distance between the longitudinal modes in such lasers is not fixed, as there is no fixed distance between the reflectors that form the laser resonator (one of the reflectors, the FBG array, is extended along the resonator). At the same time, the width of each oscillation line was extremely small—it could be seen that it was below 19 kHz (see Figure 6d). As the width of the oscillation was defined by the spectral width of each interfering longitudinal mode, we could conclude that the width of each longitudinal mode generated in our laser was below 19 kHz.

For comparison, Figure 7 shows the oscillograms of the laser with a 3 m long ytterbium-doped fiber. It can be seen that, in this case, there were no time intervals where continuous generation was observed, and the picture is noise-like, with a large number of beats at different frequencies, which indicates the generation of a large number of modes.

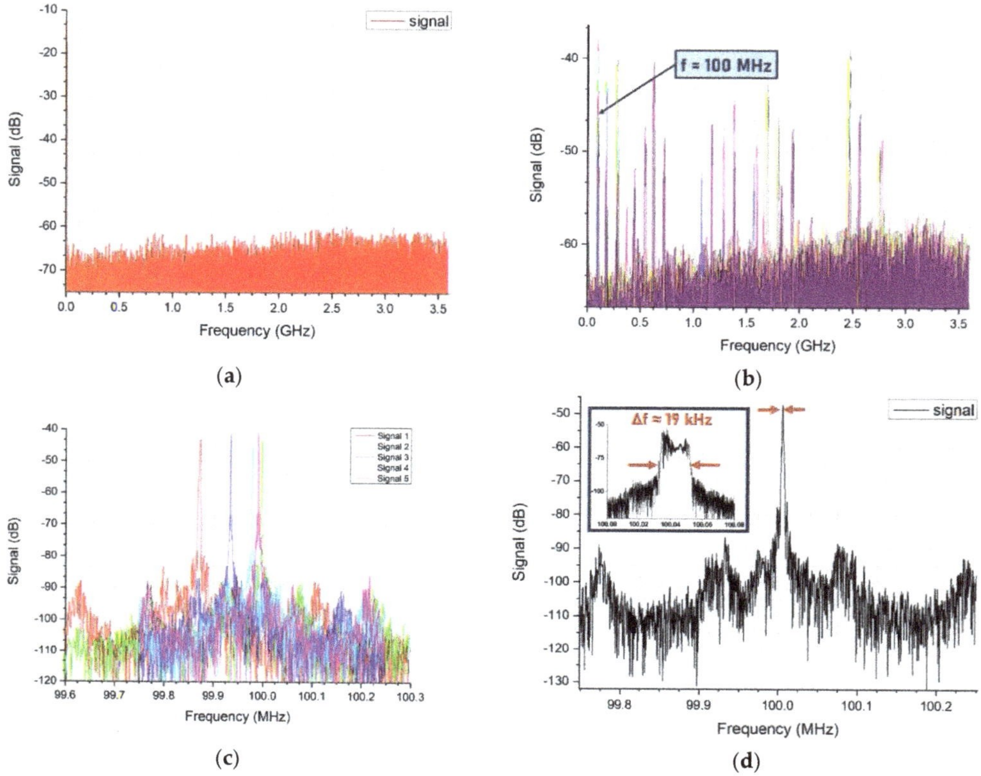

Figure 6. Radiofrequency spectra of the signal at the output of the master oscillator with YDF-FBG-PM length of 0.9 m: (**a**) Regime 1; (**b**) Regime 2 (different color corresponds to spectrums measured at different time); (**c**) variation of the position of the 100 MHz oscillation line; and (**d**) measurement of the 100 MHz oscillation bandwidth (inset—100 MHz oscillation spectrum measured in a smaller frequency range).

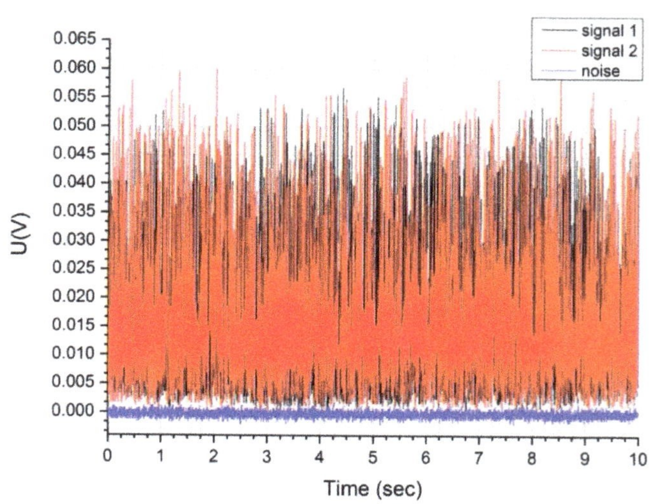

Figure 7. Graphs of signals obtained by an oscilloscope with the YDF-FBG-PM length of 3 m.

Finally, to amplify the signal, a single-stage, all-fiber, single-mode amplifier was assembled, similar to that developed in [17]. The scheme of the amplifier is shown in the inset to Figure 3, and consisted of a multi-mode pump and single-mode signal combiner (PC), multi-mode pump diode at 915 nm, YDF-DC-13/80-PM fiber available from FORC-Photonics, similar to the one developed in [17], and cladding pump striper (CPS) based on a single-mode, polarization-maintaining fiber with a core diameter of 10 μm. In the assembled amplifier, the dependence of the output power of the amplified signal on the pump power, as well as the amplified spontaneous emissions (ASE) in the wavelength regions of 976 nm and 1030 nm, was studied; the results are shown in Figure 8. An output signal power of more than 1 W was obtained, and the resulting amplification efficiency was 5.1% (it was below 1, reported in [17], due to a non-optimal signal wavelength and additional splice loss at the output of the Yb-doped fiber). It should be noted, however, that, at the maximum power, a significant level of enhanced spontaneous luminescence appeared in the spectral region near 976 nm. For this reason, the output power at the output of the first stage should be limited to 500 mW, and to achieve higher power, an additional amplification stage with a spectral filter at the input should be used—the received signal power at the output of the first stage should be sufficient for the efficient operation of a power amplifier based on special designs of ytterbium-doped fibers [12,14]. In addition, it should be noted that, to maintain the oscillation regimes described above, it was necessary to use a double-stage isolator between the master oscillator and amplifier. Otherwise, no long, single-longitudinal-mode operation regime (Regime 1) was observed—possibly due to the interference of the ASE from the amplifier and a weak signal reflected from the YDF-FBG-PM.

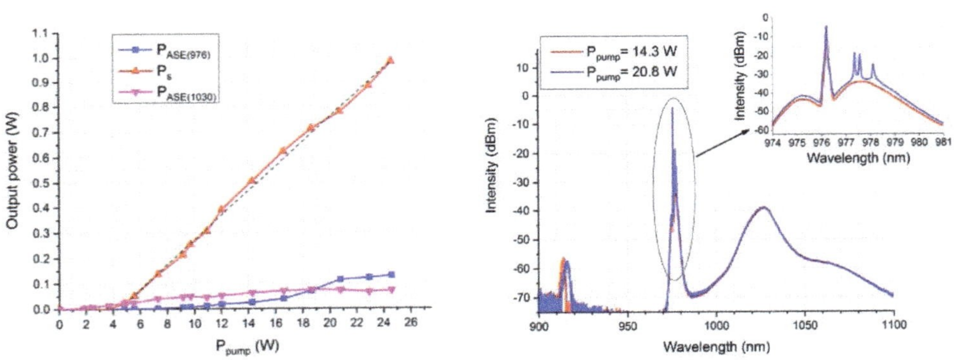

(a) (b)

Figure 8. (**a**) Plot of the pump power (P_{pump}) versus the power of amplified signal (P_s), power of ASE near 976 ($P_{ASE(976)}$), and power of ASE near 1030 ($P_{ASE(1030)}$) at the output of the amplifier (the dashed line is the result of a linear approximation of power efficiency); and (**b**) spectrum of the output signal for different level of pump power (P_{pump}).

4. Discussion

It should be noted that, in general, the observed regimes of the operation of the implemented laser with random distributed feedback were quite expected. Indeed, since the distance between the FBGs recorded in the ytterbium fiber could vary with temperature and tension, the fine structure of the reflection spectrum of the Bragg grating array could change over time due to temperature and mechanical fluctuations. As a consequence, the laser could switch the generation between different modes of the resonator, and, in between, the simultaneous generation of several modes. Apparently, this behavior is a feature of lasers with random feedback, since it is impossible to completely eliminate the effect of thermal fluctuations on a 0.9 m long fiber segment. The distance between the modes (about

200 MHz) led to the fact that switching occurred even with small changes in the external conditions. At the same time, it should be noted that, even in the absence of special efforts to thermally stabilize the laser (as was the case in our study), quite long time intervals of up to 10 s were observed when the laser operated presumably in the single-frequency mode. It also needs to be emphasized that the optical spectrum remained extremely narrow throughout the entire laser operation time—its position changed and the spectral width was below the resolution of the spectrum analyzer (0.02 nm) all the time, which makes the developed laser promising for most applications, as indicated in the introduction to this article. We expect that an improvement in the thermal stabilization of the developed laser could allow for a significantly longer (or even continuous wave) operation of the laser in the single-frequency regime.

Further optimization of the scheme is possible by increasing the Yb concentration in the core of the YDF-FBG-PM fiber. It would allow one to increase the pump-to-signal conversion efficiency of the designed laser and allow an increase in the output power for the fixed pump power. Alternatively, it would also make it possible to shorten the cavity length, increase the spacing between the longitudinal modes, and, in this way, increase the stability of the laser (would increase the time of the single-frequency laser operation, even without precise thermal stabilization).

It is also interesting to note that lasers with a long cavity, which generate random laser radiation, could also be interesting in some applications [26]. Moreover, lasers with a longer cavity could operate with a much higher efficiency.

5. Conclusions

With the optimal resonator configuration, stable generation was obtained at a wavelength of 976.2 nm with a spectral linewidth less than the resolution of the spectrum analyzer (0.02 nm), an output power of about 4 mW, and a polarization extinction ratio of more than 18 dB. An analysis of the time dependence of the output signal using a high-speed oscilloscope showed that, in the developed laser, it was possible to sufficiently realize long-time intervals (up to 10 s), during which, the laser supposedly operated in a single-frequency mode. The study of the dependence of the output power on the time using a fast oscilloscope made it possible to establish the operating mode of the laser—the generation of presumably single-frequency radiation and the switching of the generation between modes over the time intervals of the order of seconds. We believe that an improvement in the thermal stabilization of such a laser could provide continuous the wave generation of single-frequency radiation in the future.

Author Contributions: Conceptualization, A.A.R. and D.S.L.; methodology, S.S.A., V.V.V., A.A.U. and M.E.L.; software, A.A.R.; investigation, D.A.D., S.S.A., V.V.V., S.M.P., D.V.R., Y.K.C., A.A.R. and A.A.U.; data curation, S.S.A., A.A.R. and M.E.L.; writing—original draft preparation, D.A.D., S.S.A. and M.E.L.; writing—review and editing, A.A.U. and M.E.L.; supervision, D.S.L.; project administration, A.A.R. and D.S.L. All authors have read and agreed to the published version of the manuscript.

Funding: This research was funded by Russian Science Foundation, grant N 22-19-00511.

Institutional Review Board Statement: Not applicable.

Informed Consent Statement: Not applicable.

Data Availability Statement: Not applicable.

Acknowledgments: The authors are very grateful to O.V. Gryaznov and M.Yu. Artemyev from Optosystems Ltd., Moscow, Russia, for helpful consulting and for supporting the fiber Bragg gratings inscription setup.

Conflicts of Interest: The authors declare no conflict of interest.

References

1. Babin, S.A.; Churkin, D.V.; Ismagulov, A.E.; Kablukov, S.I.; Nikulin, M.A. Single frequency single polarization DFB fiber laser. *Laser Phys. Lett.* **2007**, *4*, 428–432. [CrossRef]
2. Tromberg, B.J.; Sepaniak, M.J.; Vo-Dinh, T.; Griffin, G.D. Fiber-optic chemical sensors for competitive binding fluoroimmunoassay. *Anal. Chem.* **1987**, *59*, 1226–1230. [CrossRef]
3. Wu, T.-C.; Yu, C.C.; Wang, H.-Y.; Lin, G.-R. Blue Laser Diode Enables Underwater Communication at 12.4 Gbps. *Sci. Rep.* **2017**, *7*, 40480. [CrossRef] [PubMed]
4. Bergh, A.A. Blue laser diode (LD) and light emitting diode (LED) applications. *Phys. Status Solidi* **2004**, *201*, 2740–2754.
5. Soh, D.B.S.; Codemard, C.; Wang, S.; Nilsson, J.; Sahu, J.; Laurell, F.; Philippov, V.; Jeong, Y.; Alegria, C.; Baek, S. A 980-nm Yb-Doped Fiber MOPA Source and Its Frequency Doubling. *IEEE Photonics Technol. Lett.* **2004**, *16*, 1032–1034. [CrossRef]
6. Laroche, M.; Bartolacci, C.; Cadier, B.; Gilles, H.; Girard, S.; Lablonde, L.; Robin, T. Generation of 520 mW pulsed blue light by frequency doubling of an all-fiberized 978 nm Yb-doped fiber laser source. *Opt. Lett.* **2011**, *36*, 3909–3911. [CrossRef] [PubMed]
7. Bouchier, A.; Lucas-Leclin, G.; Georges, P.; Maillard, J.M. Frequency doubling of an efficient continuous wave single-mode Yb-doped fiber laser at 978 nm in a periodically-poled MgO:LiNbO3 waveguide. *Opt. Express* **2005**, *13*, 6974–6979. [CrossRef]
8. Burkley, Z.; Brandt, A.D.; Rasor, C.; Cooper, S.F.; Yost, D.C. Highly coherent, watt-level deep-UV radiation via a frequency-quadrupled Yb-fiber laser system. *Appl. Otics* **2019**, *58*, 1657–1661. [CrossRef]
9. Vasilev, S.A.; Medvedkov, O.I.; Korolev, I.G.E.; Bozhkov, A.S.; Kurkov, A.S.; Dianov, E.M. Fibre gratings and their applications. *Quantum Electron.* **2005**, *35*, 1085. [CrossRef]
10. Seo, J.-H.; Park, J.H.; Kim, S.-I.; Park, B.J.; Ma, Z.; Choi, J.; Ju, B.-K. Nanopatterning by Laser Interference Lithography: Applications to Optical Devices. *J. Nanosci. Nanotechnol.* **2014**, *14*, 1521–1532. [CrossRef]
11. Meier, T.; Willke, B.; Danzmann, K. Continuous-wave single-frequency 532 nm laser source emitting 130 W into the fundamental transversal mode. *Opt. Lett.* **2010**, *35*, 3742–3744. [CrossRef] [PubMed]
12. Matniyaz, T.; Li, W.; Kalichevsky-Dong, M.; Hawkins, T.W.; Parsons, J.; Gu, G.; Dong, L. Highly efficient cladding-pumped single-mode three-level Yb all-solid photonic bandgap fiber lasers. *Opt. Lett.* **2019**, *44*, 807–810. [CrossRef] [PubMed]
13. Li, W.; Matniyaz, T.; Gafsi, S.; Kalichevsky-Dong, M.T.; Hawkins, T.W.; Parsons, J.; Gu, G.; Dong, L. 151W monolithic diffraction-limited Yb-doped photonic bandgap fiber laser at ~978nm. *Opt. Express* **2019**, *27*, 24972. [CrossRef] [PubMed]
14. Kotov, L.; Temyanko, V.; Aleshkina, S.; Bubnov, M.; Lipatov, D.; Likhachev, M. Efficient single-mode 976 nm amplifier based on a 45 micron outer diameter Yb-doped fiber. *Opt. Lett.* **2020**, *45*, 4292–4295. [CrossRef]
15. Boullet, J.; Zaouter, Y.; Desmarchelier, R.; Cazaux, M.; Salin, F.; Saby, J.; Bello-Doua, R.; Cormier, E. High power ytterbium-doped rod-type three-level photonic crystal fiber laser. *Opt. Express* **2008**, *16*, 17891. [CrossRef]
16. Valero, N.; Feral, C.; Lhermite, J.; Petit, S.; Royon, R.; Bardin, Y.-V.; Goeppner, M.; Dixneuf, C.; Guiraud, G.; Proulx, A.; et al. 39 W narrow spectral linewidth monolithic ytterbium-doped fiber MOPA system operating at 976 nm. *Cormier. Opt. Lett.* **2020**, *45*, 1495. [CrossRef]
17. Aleshkina, S.S.; Lipatov, D.S.; Kochergina, T.A.; Velmiskin, V.V.; Temyanko, V.L.; Kotov, L.V.; Bardina, T.L.; Bubnov, M.M.; Guryanov, A.N.; Likhachev, M.E. All-fibre single-mode small-signal amplifier operating near 0.976 μm. *Quantum Electron.* **2019**, *49*, 919. [CrossRef]
18. Röser, F.; Jauregui, C.; Limpert, J.; Tünnermann, A. 94 W 980 nm high brightness Yb-doped fiber laser. *Opt. Express* **2008**, *16*, 17310–17318. [CrossRef]
19. Vahala, K.J.; Park, N.; Dawson, J.; Newkirk, M.; Sanders, S. Semiconductor lasers and fiber lasers for fiber-optic telecommunications. *Fiber Integr. Opt.* **1992**, *11*, 221–234. [CrossRef]
20. Mo, S.; Huang, X.; Xu, S.; Li, C.; Yang, C.; Feng, Z.; Zhang, W.; Chen, D.; Yang, Z. 600-Hz linewidth short-linear-cavity fiber laser. *Opt. Lett.* **2014**, *39*, 5818–5821. [CrossRef]
21. Butov, O.V.; Rybaltovsky, A.A.; Bazakutsa, A.P.; Golant, K.M.; Vyatkin, M.Y.; Popov, S.M.; Chamorovskiy, Y.K. Chamorovskiy, "1030 nm Yb3+ distributed feedback short cavity silica-based fiber laser. *J. Opt. Soc. Am. B* **2017**, *34*, A43–A48. [CrossRef]
22. Bazakutsa, A.P.; Rybaltovsky, A.A.; Belkin, M.E.; Lipatov, D.S.; Lobanov, A.S.; Abramov, A.N.; Butov, O.V. Highly-photosensitive Er/Yb-codoped fiber for single-frequency continuous-wave fiber lasers with a short cavity for telecom applications. *Opt. Mater.* **2023**, *138*, 113669. [CrossRef]
23. Zhu, X.; Shi, W.; Zong, J.; Nguyen, D.; Norwood, R.A.; Chavez-Pirson, A.; Peyghambarian, N. 976 nm single-frequency distributed Bragg reflector fiber laser. *Opt. Lett.* **2012**, *37*, 4167–4169. [CrossRef] [PubMed]
24. Zhu, X.; Shi, W.; Zong, J.; Nguyen, D.; Norwood, R.A.; Chavez-Pirson, A.; Peyghambarian, N. Single-frequency ytter-bium-doped fiber laser at 976 nm. *Technol. Syst. Appl.* **2013**, *8601*, 86010X. [CrossRef]
25. Turitsyn, S.K.; Babin, S.A.; Churkin, D.V.; Vatnik, I.D.; Nikulin, M.; Podivilov, E.V. Random distributed feedback fibre lasers. *Phys. Rep.* **2014**, *542*, 133–193. [CrossRef]
26. Wu, H.; Xiong, J.; Han, B.; Wang, Z.; Zhang, W.; Jia, X.; Liang, H. Ultra-high speed random bit generation based on Rayleigh feedback assisted ytterbi-um-doped random fiber laser. *Sci. China Technol. Sci.* **2021**, *64*, 1295–1301. [CrossRef]
27. Skvortsov, M.I.; Abdullina, S.R.; Vlasov, A.A.; Zlobina, E.A.; Lobach, I.A.; Terentyev, V.S.; Babin, S.A. FBG array-based random distributed feedback Raman fibre laser. *Quantum Electr.* **2017**, *47*, 696–700. [CrossRef]
28. Zlobina, E.A.; Kablukov, S.I.; Babin, S.A. Generation of linearly polarised light near 1.4 μm in a cascaded, random distributed feedback Raman laser. *Quantum Electron.* **2017**, *46*, 1102–1105. [CrossRef]

29. Popov, S.; Rybaltovsky, A.; Bazakutsa, A.; Smirnov, A.; Ryakhovskiy, D.; Voloshin, V.; Kolosovskii, A.; Vorob'ev, I.; Isaev, V.; Chamorovskiy, Y.; et al. High efficient random laser with cavity based on the erbium-doped ger-manophosphosilicate artificial Rayleigh fiber. *Photonics* **2023**, *10*, 748. [CrossRef]
30. Popov, S.; Butov, O.; Bazakutsa, A.; Vyatkin, M.; Chamorovskii, Y.; Fotiadi, A. Random lasing in a short Er-doped artificial Rayleigh fiber. *Results Phys.* **2020**, *16*, 102868. [CrossRef]
31. Skvortsov, M.; Abdullina, S.; Wolf, A.; Dostovalov, A.; Churin, A.; Egorova, O.; Semjonov, S.; Proskurina, K.; Babin, S. Single-frequency erbium-doped fibre laser with random distributed feedback based on disordered structures produced by femtosecond laser radiation. *Quantum Electron.* **2021**, *51*, 1051–1055. [CrossRef]
32. Rybaltovsky, A.; Popov, S.; Ryakhovskiy, D.; Abramov, A.; Umnikov, A.; Medvedkov, O.; Voloshin, V.; Kolosovskii, A.; Vorob'ev, I.; Chamorovskiy, Y.; et al. Random Laser Based on Ytterbium Doped Fiber with a Bragg Grating Array as the Source of Continuous-Wave 976 nm Wavelength Radiation. *Photonics* **2022**, *9*, 840. [CrossRef]
33. Guan, W.; Marciante, J.R. Dual-Frequency Operation in a Short-Cavity Ytterbium-Doped Fiber Laser. *IEEE Photonics Technol. Lett.* **2007**, *19*, 261–263. [CrossRef]

photonics

MDPI

Article

Numerical Simulation of High-Power Optical Amplifiers at 2.3 μm Based on a Special Multicore Fiber

Elena A. Anashkina *, Alexey V. Andrianov and Alexander G. Litvak

A.V. Gaponov-Grekhov Institute of Applied Physics of the Russian Academy of Sciences, 46 Ulyanov Street, 603950 Nizhny Novgorod, Russia

* Correspondence: elena.anashkina@ipfran.ru

Abstract: The development of high-power laser sources at 2.3 μm is highly demanded for remote sensing and other applications. However, this wavelength is poorly covered by present-day lasers. To obtain 100 W class high-power radiation at 2.3 μm, we propose to use simultaneously cascade laser amplification at 2 and 2.3 μm with a commercially available diode pump at 793 nm and stimulated Raman scattering between the amplified signal waves in a special zinc–tellurite multicore fiber with ten trivalent-thulium-ion-doped cores arranged in a ring. We demonstrate numerically that the use of an out-of-phase supermode (with spatial phases differing by π in neighboring cores) can provide up to 50% efficiency conversion from the 793 nm pump to the 2.3 μm wave.

Keywords: multicore fiber; tellurite glass fiber; stimulated Raman scattering (SRS); Raman gain; trivalent thulium ions; laser amplification

Citation: Anashkina, E.A.; Andrianov, A.V.; Litvak, A.G. Numerical Simulation of High-Power Optical Amplifiers at 2.3 μm Based on a Special Multicore Fiber. *Photonics* **2023**, *10*, 711. https://doi.org/10.3390/photonics10070711

Received: 17 May 2023
Revised: 11 June 2023
Accepted: 17 June 2023
Published: 21 June 2023

1. Introduction

In recent years, the development of coherent light sources at a wavelength of about 2.3–2.5 μm has been attracting increasing interest [1–7]. One of the main reasons is that the earth's atmosphere is highly transparent in this spectral region. Hazardous gases such as HF, CH_4, and CO have pronounced absorption lines near 2.3 μm, which opens up an opportunity for using 2.3 μm lasers for remote sensing. In addition, laser sources at this wavelength are in demand for LiDARs and other applications. Many real applications require high power (tens of watts and even higher), which is a challenge for modern laser systems, but works in this direction are actively underway based on solid-state lasers [1–3] and fiber lasers and amplifiers [4–7]. Bulk lasers can provide higher output powers on average, but fiber lasers are generally characterized by high beam quality, alignment-free operation, and good heat dissipation. In addition, some fiber laser systems are capable of producing high average power too (~100 W and higher).

A very promising way to develop laser sources operating beyond 2.1 μm is based on using tellurite (tellurium dioxide-based) glass fibers [8]. Silica glasses commonly used for near-IR fiber devices have large losses beyond 2.1 μm, so their use is very limited at longer wavelengths (although not excluded [9]). Tellurite glasses are transparent in the range of ~0.5–5 μm, have excellent optical and physicochemical properties [8,10–13] and fairly high damage thresholds (i.e., they can resist high peak intensities [14]), can be doped with high concentrations of active ions [15], and are produced with low background losses [8]. Lasers at 2.3 μm based on tellurite fibers doped with trivalent thulium ions have been successfully demonstrated [5–7], but their experimental powers still do not exceed several hundred mW [7]. Another promising approach to generate light at 2.3 μm is to use stimulated Raman scattering (SRS) in undoped tellurite fibers pumped at 2 μm. This approach was theoretically explored in [16], where a continuous wave (CW) Raman laser with a power of ~160 W at a wavelength of 2.35 μm based on a single-mode fiber with a pump power of 300 W at 2 μm was demonstrated numerically. However, single-mode fiber lasers at 2 μm with a power of 300 W are non-standard and rather complex devices, manufactured only in individual scientific laboratories. In addition, it remains an open question whether a single-mode tellurite fiber can resist such CW powers.

In this work, for obtaining high-power radiation at 2.3 μm in a Tm-doped tellurite multicore fiber amplifier, we propose to simultaneously use a unique combination of cascade lasing at two radiative transitions together with SRS processes. This is inherent specifically for zinc–tellurite (TZ, TeO_2–ZnO-based) glasses doped with trivalent thulium ions, which, in principle, makes it possible to obtain the efficiency of conversion of the pump at 793 nm to a signal at 2.3 μm above the Stokes limit (0.793 μm/2.3 μm = 34.5%). A simplified diagram of laser levels of trivalent thulium in a TZ glass matrix is shown in Figure 1a [15]. Previously, it was shown that the cascade scheme of dual-band operation at the $^3H_4 \to {}^3H_5$ and $^3F_4 \to {}^3H_6$ transitions with a pump wave at 793 nm can significantly increase the efficiency for a wave at 2.3 μm compared to operation only at the $^3H_4 \to {}^3H_5$ transition, since it helps to overcome the bottleneck problem related to a long lifetime of $^3F_4$ [5]. It is also known that Raman gain is very large for TZ glasses, with the maximum at about 750 cm^{-1} (22.5 THz), which is ~60 times higher than the maximum for silica glass [15]. Therefore, additional amplification of a signal at 2.3 μm due to the SRS of the wave at ~2 μm is also possible. The scheme of the SRS process is shown in Figure 1b. In this case, for the Raman amplification at 2.3 μm, the wavelength of 1.96 μm displaced by 22.5 THz is most suitable. The Raman gain of the TZ glass in comparison with the Raman gain of the silica glass is plotted in Figure 1c based on the data presented in [17]. Note that in [5], the parameters of the system were such that the SRS processes for them were negligible.

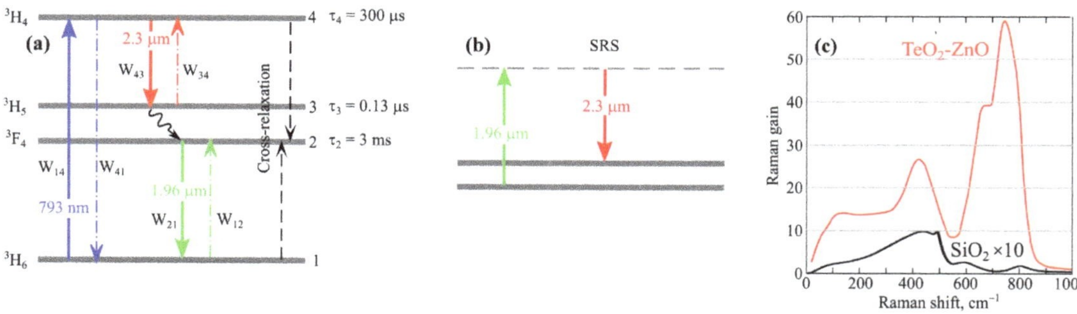

Figure 1. (**a**) Simplified diagram of the energy levels of trivalent thulium ions in the TZ glass matrix. W_{jk} are stimulated rates from level j to level k, and τ_j is the total lifetime of level j. (**b**) Schematic diagram of the SRS process. (**c**) Raman gain spectra of TZ glass (red) compared to silica glass (black; magnification by a factor of 10).

The concept of a master-oscillator power amplifier (MOPA) considered in this paper is often used for high-power laser systems [18,19]. At the same time, special fiber designs, e.g., multimode large mode area (LMA) fibers operating in quasi-single mode, are frequently developed for high-power amplifiers [18,20,21].

It is also possible to use taper fibers with a thin single-mode core at the input and a thick core at the out-put [22,23]. With adiabatic signal amplification, the output beam quality can be very high, while the scattering into higher modes is rather small. One can also use multicore fibers with N cores when each core is a single-mode one [24–26]. In the case of evanescent-field coupling, there are N supermodes in the system [25,27]. It was demonstrated that the use of an out-of-phase supermode of a multicore fiber with a symmetrical arrangement of axially symmetric cores, for which the spatial phases in neighboring cores differ by π, has a number of advantages over a "common" in-phase supermode, for which the phases are the same in all cores [28–31]. This issue was dis-cussed in detail in [28–31], here we only note that the in-phase mode is unstable at high powers, while the out-of-phase mode is stable. Moreover, it has been predicted theoretically that the self-focusing limit for the beam power (for continuous medium) can be overcome using the out-of-phase mode, while this is impossible for the in-phase mode due to its instability [28,29]. As shown experimentally and numerically, with selective excitation of the out-of-phase supermode during amplification, the mode composition at the output does not deteriorate [30]. Moreover, the out-of-phase mode overlaps better with the doped core, so it has a higher gain (this statement will be demonstrated below). Multicore fibers can be fabricated from various glasses, including tellurite ones. A rare-earth ion-doped multicore fiber was experimentally reported in [32], and a passive seven-core tellurite fiber was demonstrated, for example, in [33]. Although the manufacture of high-quality multicore tellurite fibers is still a challenge today, in principle, technologies can be improved to the required level.

Here, we propose and numerically investigate high-power amplifiers at a wavelength of 2.3 μm based on a special double-clad TZ fiber with N = 10 step-index cores doped with trivalent thulium ions symmetrically arranged along the ring. A multicore fiber geometry leads to a decrease in peak intensities of laser beams relative to a single-mode fiber geometry and helps to prevent potential fiber damage for high-power laser amplifiers. It will be demonstrated that the use of the out-of-phase supermode is more efficient than the in-phase supermode. We propose to use a low-power seed signal at 2.3 μm and an additional low-power seed signal at ~2 μm with a standard 300 W diode pump at 793 nm. Two seeding low-power waves at 2.3 μm and 1.96 μm (with a power of about 1 mW) can be generated simultaneously in laser elements based on various matrices doped with trivalent thulium ions [5–7]. Note that at present, relatively cheap commercially available laser diodes provide powers up to several hundred watts at a wavelength of 793 nm. In this case, the efficiency of converting the pump power to a wave at 2.3 μm, as will be shown below, can reach 50% due to the combination of laser and SRS processes. The scheme of a cascade laser amplifier, with one wave used for additional amplification of another due to SRS, has not been studied before to the best of our knowledge.

2. Materials and Methods

We considered a special design of a double-clad TZ glass multicore fiber shown in Figure 2a. Ten step-index cores doped with trivalent thulium ions are arranged in a ring. The diameter of each core is d = 7 μm, and the distance between the centers of neighboring cores is $L = 1.2 \times d$ = 8.4 μm. The refractive index of the first undoped cladding is chosen to be 0.1 less than the refractive index of the core glass (Figure 2b), which can be achieved by a slight change in the content of zinc oxide in TZ glass compositions [5]. The parameters of each core are chosen so that it is single mode at signal wavelengths. A diode pump at 793 nm is launched into the first undoped cladding, which has a diameter of D = 200 μm. This diameter D coincides with the typical diameter of the fiber core, which outputs a diode pump with a power of hundreds of watts. In this case, the numerical aperture between undoped cladding and cladding 2 of the TZ multicore fiber should correspond to the numerical aperture of the fiber that outputs the pump (typically NA = 0.22). The multicore fiber also has the second cladding (its diameter is not important in the calculations).

We found supermode structures for the special multicore TZ fiber using finite element modeling, as in our previous work [34]. Figure 3 shows examples of all ten supermodes calculated at a wavelength of 1.96 μm. Mode 1 (in-phase supermode) has the highest effective refractive index, and mode 10 (out-of-phase supermode) has the lowest effective refractive index. The pair of modes 2 and 3 is degenerate in the effective refractive index. The pairs of modes 4, 5 as well as 6, 7, and 8, 9 are also degenerate. The in-phase and out-of-phase supermodes have the most symmetrical structures. For them, $|E_x|$ distributions are the same for each core. For other modes, $|E_x|$ distributions are not the same in different cores; moreover, in some cores, the fields are close to zero or small. Therefore, for modes 2–9, the gain coefficients provided by trivalent thulium ions are obviously lower due to the smaller overlap with the doped region than for modes 1 and 10. Moreover, the preparation of seed signals for amplifiers in modes 2–9 is more difficult than in modes 1 and 10. In-phase and out-of-phase supermodes can be prepared using, for example, a special beam shaper (systems based on spatial light modulators are commonly used for such purposes) [30,35]. Further, we consider only in-phase and out-of-phase modes.

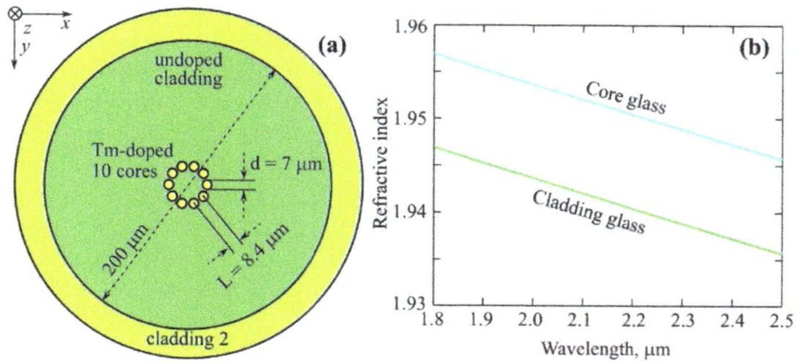

Figure 2. (**a**) Cross section of the proposed multicore TZ glass fiber with trivalent thulium-doped cores. (**b**) Wavelength-dependent refractive indexes for core and cladding TZ glasses.

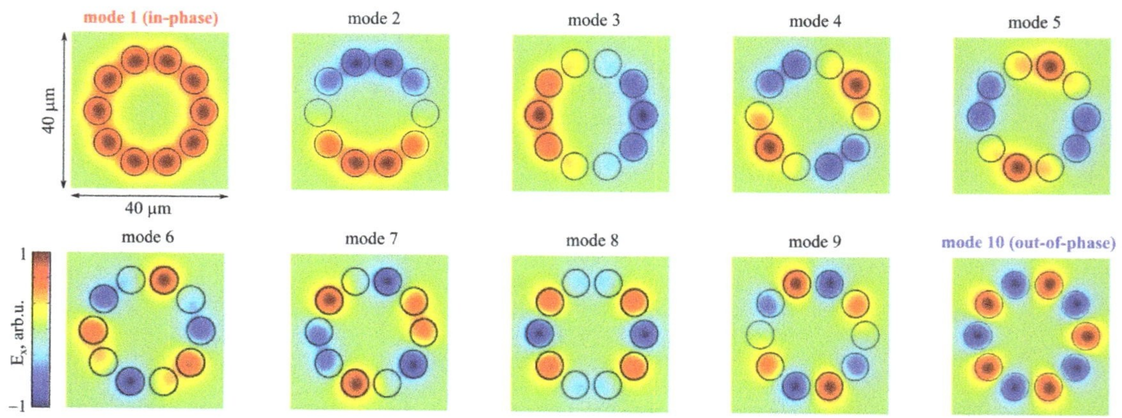

Figure 3. Calculated distributions of x-components of electric fields for supermodes at 1.96 μm. The lowest supermode (mode 1) is in-phase one. The highest supermode (mode 10) is out-of-phase one.

We calculated effective mode areas for in-phase and out-of-phase supermodes as functions of wavelength. The results are plotted in Figure 4. The mode spatial structures at 1.96 μm and 2.3 μm which are of primary interest are shown in the insets in Figure 4.

Effective supermode areas are calculated as

$$A_{eff} = \frac{\left(\int S_z dxdy\right)^2}{S_z^2 dxdy}, \tag{1}$$

where S_z is the z-component of the Poynting vector.

A schematic diagram of the considered multicore fiber laser amplifier seeded by a dual-wavelength source at 1.96 μm and 2.3 μm with a high-power diode pump at 793 nm is shown in Figure 5. To excite a required supermode, a beam shaper, for example, based on a spatial light modulator can be used [30].

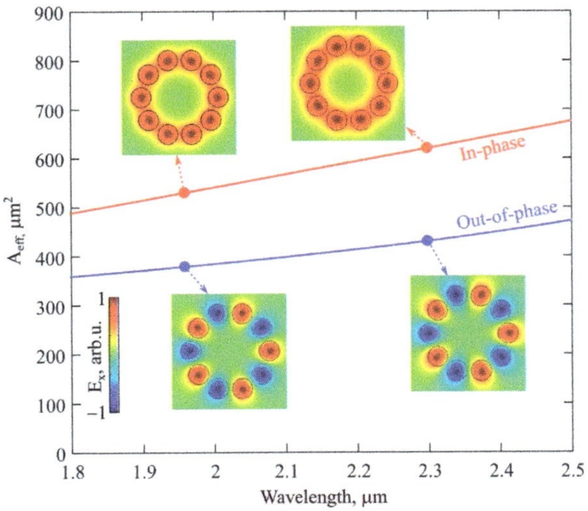

Figure 4. Wavelength-dependent effective mode areas for in-phase and out-of-phase supermodes. Insets show the corresponding distributions of x-components of electric fields E_x at 1.96 μm and 2.3 μm.

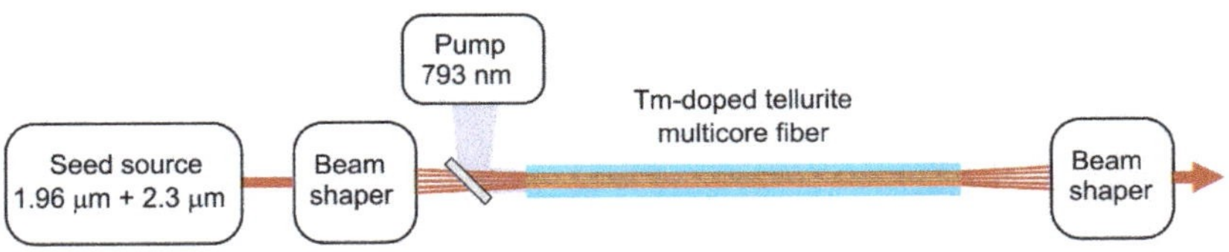

Figure 5. Schematic diagram of the laser amplifier based on Tm-doped tellurite multicore fiber.

To simulate a dual-wavelength laser amplifier with allowance for the SRS process between waves at 1.96 and 2.3 μm, we generalized the system of rate equations and the power evolution equations presented in [5,15]. The SRS was considered in a standard way [27]. We analyzed CW waves, but this model can be valid for describing pulsed amplifiers with signal durations noticeable longer than the total lifetime of level $^3F_4$. In this case, all powers refer to peak values. This approach may be justified when average powers should be reduced. The rate equations for the population densities n_1, n_2, n_3, and n_4 (normalized to the concentration of trivalent thulium ions N_{Tm}) in the steady state are [5,15]:

$$n_1 + n_2 + n_3 + n_4 = 1 \quad (2)$$

$$\frac{\partial n_4}{\partial t} = W_{14}n_1 + W_{34}n_3 - \left(W_{41} + W_{43} + \frac{1}{\tau_4}\right)n_4 - K_{CR}n_4n_1 = 0 \quad (3)$$

$$\frac{\partial n_3}{\partial t} = -\left(W_{34} + \frac{1}{\tau_3}\right)n_3 + \left(W_{43} + \frac{\beta_{43}}{\tau_4^R} + \frac{1}{\tau_4^{NR}}\right)n_4 = 0 \quad (4)$$

$$\frac{\partial n_2}{\partial t} = W_{12}n_1 - \left(W_{21} + \frac{1}{\tau_2}\right)n_2 + \frac{n_3}{\tau_3} + \frac{\beta_{42}}{\tau_4^R}n_4 + 2K_{CR}n_4n_1 = 0, \quad (5)$$

where t is time, τ_j is the total lifetime of level j, $\tau_4{}^R$ and $\tau_4{}^{NR}$ are the radiative and non-radiative lifetimes of level 4, W_{jk} are the stimulated rates from level j to level k, β_{4j} is the branching ratio from level 4 to level j, and K_{CR} is the cross-relaxation rate (see the scheme in Figure 1a) [5,15]. Stimulated absorption and emission rates for a pump wave are:

$$W_{14,41} = \frac{\Gamma_p \lambda_p \sigma_{14,41} P_p}{hcA_{doped}} \tag{6}$$

where c is the speed of light, h is the Planck's constant, σ_{14} and σ_{41} are the absorption and emission cross sections at a pump wavelength λ_p, P_p is the wave power, A_{doped} is the area of doped cores, and Γ_p is the overlap integral of the pump intensity distribution with the doped regions estimated as $\Gamma_p = A_{doped}/(\pi D^2/4)$. Stimulated emission and absorption rates for signal waves at λ_1 = 1.96 μm and λ_2 = 2.3 μm with powers P_1 and P_2, respectively, are:

$$W_{21,12} = \frac{\Gamma_1 \lambda_1 \sigma_{21,12} P_1}{hcA_{doped}} \tag{7}$$

$$W_{43,34} = \frac{\Gamma_2 \lambda_2 \sigma_{43,34} P_2}{hcA_{doped}}, \tag{8}$$

where $\Gamma_{1,2}$ are the overlap integrals of the signal waves with the doped regions.

The SRS-modified equations for the power evolution along the z-axis are:

$$\frac{dP_p}{dz} = -\Gamma_p N_{Tm}(\sigma_{14}n_1 - \sigma_{41}n_4)P_p - \alpha_p P_p \tag{9}$$

$$\frac{dP_1}{dz} = \Gamma_1 N_{Tm}(\sigma_{21}n_2 - \sigma_{12}n_1)P_1 - g_{R12}\frac{\lambda_2}{\lambda_1}P_2 P_1 - \alpha_1 P_1 \tag{10}$$

$$\frac{dP_2}{dz} = \Gamma_2 N_{Tm}(\sigma_{43}n_4 - \sigma_{34}n_3)P_2 + g_{R12}P_1 P_2 - \alpha_2 P_2 \tag{11}$$

where α_1, α_2, and α_p are the linear losses at wavelengths λ_1, λ_2, and λ_p, respectively; g_{R12} is the Raman gain coefficient for the TZ fiber calculated as

$$g_{R12} = \frac{g_R(\mathrm{TZ@1\mu m}) \times (1\mu\mathrm{m}/\lambda_1)}{\left(A_{eff1} + A_{eff2}\right)/2} \tag{12}$$

Here, g_R(TZ@1μm) is the maximum Raman gain for bulk TZ glass for the pump wave at 1 μm [17,27]. For most of the simulations, we set optical losses of 0.3 dB/km for laser waves, which is easily achievable for modern single-core tellurite fibers and, in principle, can be achieved for multicore tellurite fibers. At the very end of the section Results, we will also analyze the impact of higher losses (up to 5 dB/m) that can be caused by light scattering at the core-cladding interfaces, bending losses, and so on.

The system of Equations (9)–(11), with allowance for the rate Equations (2)–(5) in the steady state, was solved numerically by the classical Runge–Kutta method (with adaptive step size). The used parameters are listed in Table 1. We neglected the spontaneous Raman scattering to the anti-Stokes wave and the second-order Raman scattering to the Stokes wave. At room temperature, the process of scattering to the anti-Stokes component is weak. Moreover, its expected wavelength of 1.7 μm is strongly absorbed by Tm ions. The 2nd order Stokes Raman wave at the expected wavelength of about 2.8 μm is strongly absorbed by hydroxyl groups in zinc–tellurite glass.

Table 1. Parameters used in simulation.

Parameter	Symbol	Value
Core diameter	d	7 μm
Distance between centers of neighboring cores	L	8.4 μm
Number of cores	N	10
Cladding diameter	D	200 μm
Pump wavelength	λ_p	793 nm
Wavelength at $^3F_4 \to {}^3H_6$ transition	λ_1	1.96 μm
Wavelength at $^3H_4 \to {}^3H_5$ transition	λ_2	2.3 μm
Effective supermode area at 1.96 μm	A_{eff1}	530 μm^2 (in-phase) 380 μm^2 (out-of-phase)
Effective supermode area at 2.3 μm	A_{eff2}	620 μm^2 (in-phase) 430 μm^2 (out-of-phase)
Overlap integral (pump with doped area)	Γ_p	0.012
Overlap integral (wave at 1.96 μm with doped area)	Γ_1	0.73 (in-phase) 0.95 (out-of-phase)
Overlap integral (wave at 2.3 μm with doped area)	Γ_2	0.62 (in-phase) 0.9 (out-of-phase)
Fiber loss at 793 nm	α_p	0.4 dB/m
Fiber loss at 1.96 μm	α_1	0.3 dB/m (Figures 6–9) 0.3–5 dB/m (Figure 10)
Fiber loss at 2.3 μm	α_2	0.3 dB/m
Raman gain of bulk TZ glass at 1 μm	g_R(TZ@1 μm)	59×10^{-13} m/W
Raman gain coefficient (for wave at 2.3 μm amplified by wave at 1.96 μm)	g_{R12}	7.3×10^{-5} (W cm)$^{-1}$ (in-phase) 5.1×10^{-5} (W cm)$^{-1}$ (out-of-phase)
Absorption cross section at $^3H_6 \to {}^3H_4$	σ_{14}	1×10^{-24} m^2
Emission cross section at $^3H_4 \to {}^3H_6$	σ_{41}	1×10^{-24} m^2
Emission cross section at $^3F_4 \to {}^3H_6$	σ_{21}	2.6×10^{-25} m^2
Absorption cross section at $^3H_6 \to {}^3F_4$	σ_{12}	1.2×10^{-26} m^2
Emission cross section at $^3H_4 \to {}^3H_5$	σ_{43}	2.6×10^{-25} m^2
Absorption cross section at $^3H_5 \to {}^3H_4$	σ_{34}	2.6×10^{-25} m^2
Total lifetime of level $^3H_6$	τ_4	300 μs
Radiative lifetime of level $^3H_6$	$\tau_4{}^R$	400 μs
Non-radiative lifetime of level $^3H_6$	$\tau_4{}^{NR}$	1.2 ms
Total (non-radiative) lifetime of level $^3H_5$	τ_3	0.13 μs
Total lifetime of level $^3F_4$	τ_2	3 ms
Branching ratio from level 4 ($^3H_6$) to level j	β_{41} β_{42} β_{43}	0.9 0.07 0.03
Cross-relaxation rate	K_{CR}	1262 s^{-1} (for $N_{Tm} = 1 \times 10^{20}$ cm^{-3}) 5678 s^{-1} (for $N_{Tm} = 2 \times 10^{20}$ cm^{-3}) 31,068 s^{-1} (for $N_{Tm} = 4 \times 10^{20}$ cm^{-3}) 164,370 s^{-1} (for $N_{Tm} = 8 \times 10^{20}$ cm^{-3})

3. Results

Let us first revisit the main idea of the paper: in high-power dual-wavelength fiber amplifiers based on a special TZ fiber doped with trivalent thulium ions, the interaction between waves at 1.96 μm and 2.3 μm due to SRS leads to a significant increase in the efficiency of power conversion from the diode pump at 793 nm to the wave at 2.3 μm. Moreover, the conversion efficiency is higher for the out-of-phase supermode than for the in-phase one because the effective area of the out-of-phase supermode is smaller than that of the in-phase one (for example, the effective mode area is 430 μm^2 for out-of-phase and 620 μm^2 for in-phase supermodes at 2.3 μm); the out-of-phase supermode is better localized near the cores. This is due to the fact that the topological structures of the field distributions of in-phase and out-of-phase modes are different. There are lines of zero field amplitude between the cores for the out-of-phase mode (due to equal amplitudes and phase difference by π at these lines), so most of its power is contained in the cores. On the contrary, there is significant field amplitude in the regions between the cores for the in-phase mode, as clearly seen in Figure 4. The overlap integrals with doped cores for the out-of-phase supermode are higher (see Table 1); hence, the laser gain coefficients are higher too. In addition, the Raman gain is also higher for the out-of-phase supermode (see Table 1).

Figure 6 (left column) shows the results of calculations performed for the out-of-phase supermode, and Figure 6 (right column) for the in-phase one. Hereinafter, the seed signal powers at 1.96 μm and at 2.3 mW are 10 mW and 1 mW, respectively. We checked that the exact values of low-power seed signals do not affect the output powers of 100-W class signals. The pump power was 200 W, and the concentration of active ions was $N_{Tm} = 4 \times 10^{20}$ cm^{-3}. We purposely switched on and off the SRS in the calculations to reveal its contribution. The results obtained for the complete model are plotted as solid curves and for the model without SRS as dashed curves. Figure 6a,b show the evolution of the wave power at 2.3 μm, and Figure 6c,d at 1.96 μm; whereas Figure 6e,f show the evolution of the pump wave power.

At short distances from the fiber input end, there is cascading laser amplification of both signal waves, and the contribution of the SRS is small. The SRS effect becomes significant when the power in the wave at 1.96 μm is sufficiently large and almost reaches its maximum. This occurs approximately when most of the pump power has been absorbed (vertical line through Figure 6a,c,e). After that, the wave at 1.96 μm ceases to be amplified. In the complete model with SRS, the power is further efficiently transferred from the wave at 1.96 μm to the wave at 2.3 μm. In the model without SRS, the powers in both signal waves reach a maximum at close points z and then decay due to losses; there is no further interaction between them. The wave at 1.96 μm decays faster since it is partially absorbed from the ground state (at the $^3H_6 \rightarrow {}^3F_4$ transition). The wave at 2.3 μm is not absorbed at the $^3H_6 \rightarrow {}^3F_4$ transition. At practically zero pump powers, the levels above $^3F_4$ are not populated; therefore, the wave at 2.3 μm cannot be absorbed from excited states. Note that the power distribution in the pump wave is practically the same both in the model with and without SRS. In the case with allowance for SRS, the maximum power at 2.3 μm is 86 W in the out-of-phase supermode and 66 W in the in-phase one. For the out-of-phase supermode, the diode pump power conversion efficiency is 43%, exceeding the Stokes limit of 34.5%. Further study and optimization allowed us to find parameters of the system for which it is possible to achieve higher values of powers and conversion efficiencies.

Next, we studied the dependence of the power in two signal waves on two variables: the diode pump power and the fiber length for out-of-phase (Figure 7, left-hand panels) and in-phase (Figure 7, right-hand panels) supermodes with and without SRS (at $N_{Tm} = 4 \times 10^{20}$ cm^{-3}). The power curves at 2.3 μm are plotted in the top row (Figure 7a–d) and at 1.96 μm in the bottom row (Figure 7e–h).

It can be seen that at pump power < 100 W, the wave at 2.3 μm is weakly amplified, while the wave at 1.96 μm is significantly enhanced. This is explained by the cross-relaxation effect, which is rather strong at the chosen concentration of active ions. Its action leads to a decrease in the population at the $^3H_4$ level and a decrease in the laser gain at the $^3H_4 \rightarrow {}^3H_5$ transition, as well as an increase in the population at the $^3F_4$ level and a significant increase in the laser gain at the $^3H_4 \rightarrow {}^3H_6$ transition. Thus, at a pump power of 100 W, the maximum wave power at 1.96 μm is 60 W (the conversion efficiency reaches 60%) in the out-of-phase mode, which significantly exceeds its Stokes limit (0.793/1.96 = 40%). In this case, the red dashed-dotted curves in the lower panels demonstrate the optimal lengths of the wave at 1.96 μm depending on the pump power (i.e., the z value at which the maximum P_1 is reached at a certain P_p). These dependences are rather weak and are determined by the pump absorption length, which changes weakly for the considered parameters. The blue-dotted curves in all plots show the optimal fiber lengths for maximizing wave power at 2.3 μm as a function of pump power (i.e., the z value at which P_2 is maximized at a certain P_p).

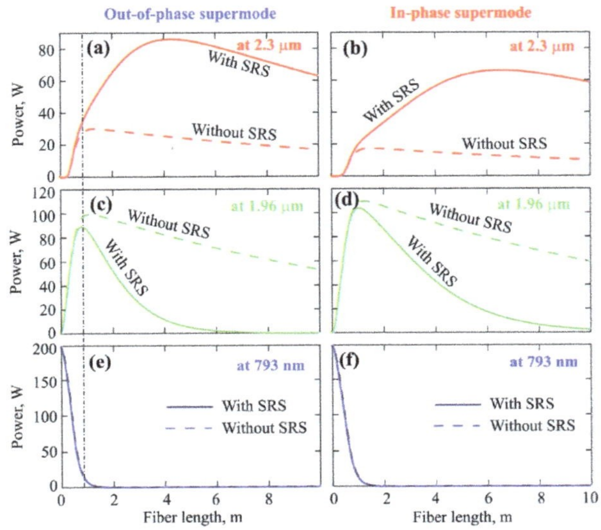

Figure 6. Powers vs. fiber length calculated with/without SRS for pump power 200 W and N_{Tm} = 4 × 10^{20} cm^{-3} for waves at 2.3 μm propagating in out-of-phase (**a**) or in-phase (**b**) supermodes; for waves at 1.96 μm in out-of-phase (**c**) or in-phase (**d**) supermodes; and for waves at 793 nm in out-of-phase (**e**) or in-phase (**f**) supermodes. The vertical dash-dotted line through Figures (**a**,**c**,**e**) shows the fiber length at which the power in the wave at 1.96 μm reaches its maximum.

In the case of switching off the SRS process, the optimal lengths are quite close for both waves since amplification occurs only due to cascade laser processes, which cease when the pump is absorbed. In the case of an allowance for SRS, the optimal lengths maximizing wave power at 2.3 μm strongly depend on the pump power. The higher the pump power, the shorter the optimal length. This has a very simple explanation: the higher the pump power, the higher the power in the wave at 1.96 μm, which means that the faster it amplifies the wave at 2.3 μm due to SRS. The optimal length for a wave at 2.3 μm is reached when the gain becomes equal to the optical loss. With the chosen TZ fiber parameters, this occurs for P_1 = 9.5 W for the out-of-phase mode and P_1 = 13.5 W for the in-phase mode (i.e., the blue-dotted curves in Figure 7e,g show the level lines corresponding to these values). In addition, as P_p increases, the population of the $^3H_6$ level increases, and the laser effects are also enhanced for the wave at 2.3 μm. At any pump power, the maximum power achieved in the wave at 2.3 μm is higher for the out-of-phase supermode than for the in-phase one.

Next, we made optimizations for the wave at 2.3 μm for various concentrations of active ions. The ratio of the power in this wave to the pump power (P_2/P_p) as a function of two variables, pump power and fiber length, is presented in Figure 8.

The following features are visible here: At low concentrations of trivalent thulium ions, amplification of the wave at 2.3 μm with a noticeable efficiency (>10%) begins at a pump power of about 10 W (Figure 8a), while with increasing concentration, a higher pump power is required (Figure 8b–d). At the highest considered value of N_{Tm}, the cross-relaxation is so strong that the laser amplification of the wave at 2.3 μm is small, and notable values of its power are achieved mainly due to the SRS process at high powers and long fiber lengths (Figure 8d). The optimum for P_2 is achieved at intermediate concentrations of N_{Tm} =24 × 10^{20} cm^{-3} at high pump power (P_p > 150 W) (Figure 8b,c). In this case, both effects are important: laser amplification and SRS. The maximum conversion efficiency reaches 50%, which exceeds the Stokes limit by 15%.

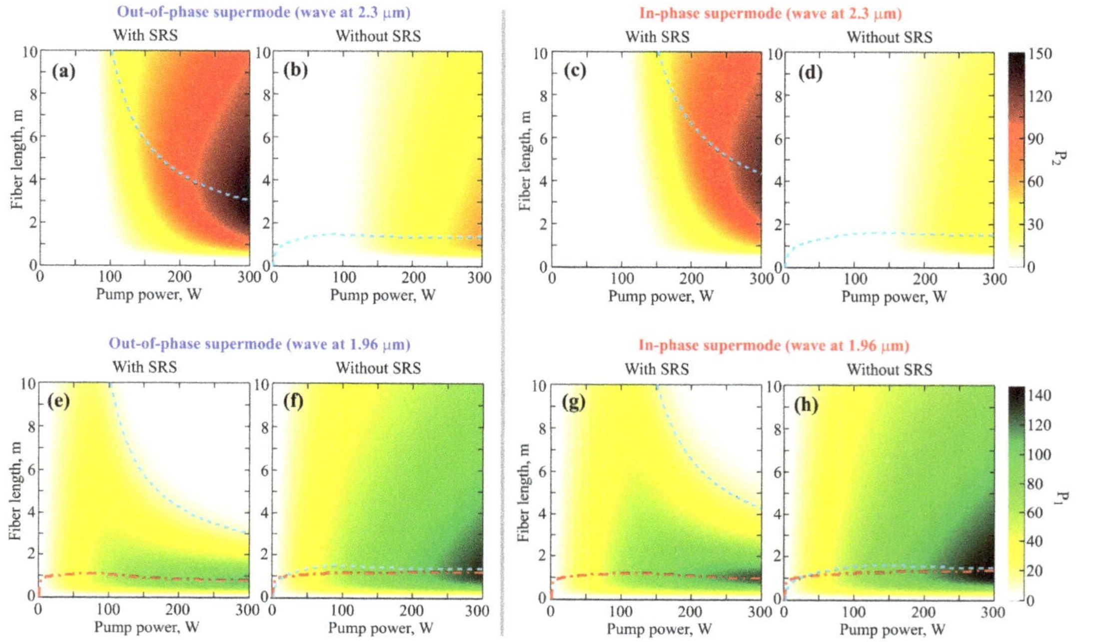

Figure 7. (**a**) Power in the wave at 2.3 μm vs. two variables: pump power and fiber length calculated for out-of-phase mode with SRS (**a**); out-of-phase mode without SRS (**b**); in-phase mode with SRS (**c**); in-phase mode without SRS (**d**). Power in the wave at 1.96 μm vs. pump power and fiber length calculated for out-of-phase mode with SRS (**e**); out-of-phase mode without SRS (**f**); in-phase mode with SRS (**g**); in-phase mode without SRS (**h**). Dotted blue curves in all panels demonstrate the optimal fiber length maximizing P_2 for certain pump power. Dash-dotted red curves in (**e**–**h**) correspond to the fiber lengths maximizing P_1 for certain pump powers.

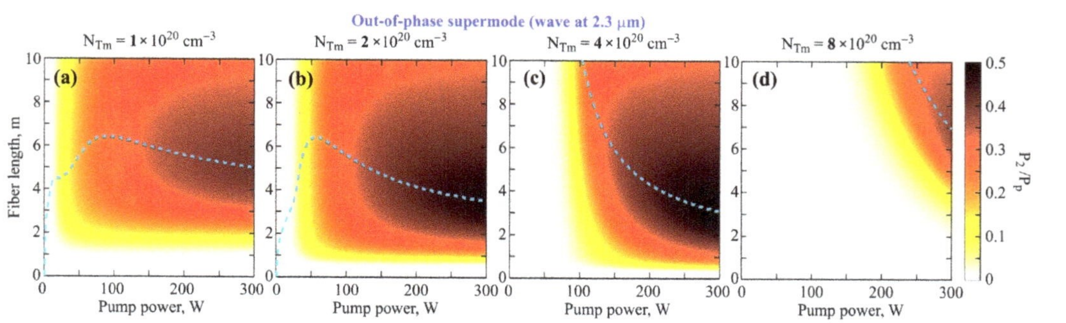

Figure 8. (**a**) Ratio of the power in the wave at 2.3 μm to the pump power (P_2/P_p) as a function of two variables: pump power and fiber length calculated for N_{Tm} = 1 × 10^{20} cm^{-3} (**a**); N_{Tm} = 2 × 10^{20} cm^{-3} (**b**); N_{Tm} = 4 × 10^{20} cm^{-3} (**c**); and N_{Tm} = 8 × 10^{20} cm^{-3} (**d**). Dotted blue curves in all panels demonstrate the optimal fiber length maximizing P_2 for certain pump power.

Next, for clarity, we plotted the dependence of the wave power at 2.3 μm on the pump power for optimal concentrations: $N_{Tm} = 2 \times 10^{20}$ cm^{-3} (Figure 9a) and $N_{Tm} = 4 \times 10^{20}$ cm^{-3} (Figure 9b). Each curve in Figure 9a,b corresponds to a certain length L = 1, 2, ... , 7 m and, in fact, reproduces the data from Figure 8b,c. It is clearly seen that at the most optimal lengths of 3–4 m and at high powers (>150 W), these dependences are close to linear functions.

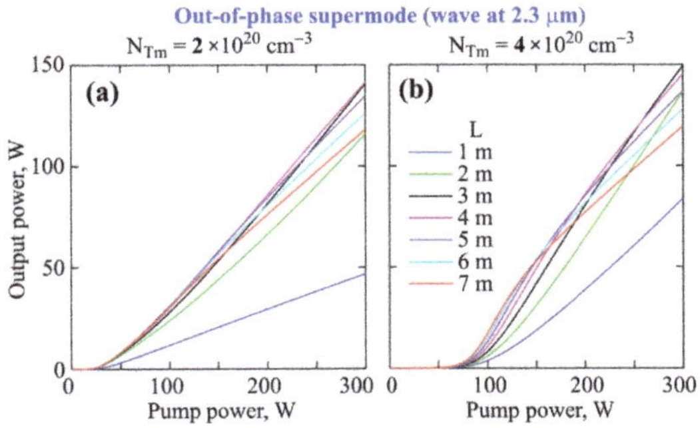

Figure 9. Power in the wave at 2.3 μm as a function of pump power for $N_{Tm} = 2 \times 10^{20}$ cm^{-3} (**a**) and $N_{Tm} = 4 \times 10^{20}$ cm^{-3} (**b**) calculated for different fiber lengths. The legend in (**b**) is shared for both panels.

Finally, we studied the effect of increased fiber losses on the quantitative characteristics of laser amplifiers. For all simulations presented above, we set optimistic losses of 0.4 dB/m at 793 nm and 0.3 dB/m at 1.96 and 2.3 μm. Here, we purposely increased losses at laser wavelengths to take into account additional imperfections (including bending losses). The simulated output powers at 2.3 μm vs. the fiber length for optimal concentration $N_{Tm} = 4 \times 10^{20}$ cm^{-3} calculated for out-of-phase supermodes with SRS and without SRS are plotted in Figure 10a,b, respectively, for different losses in the 0.3–5 dB/m range. It is not surprising that the higher the loss, the lower the output power and the shorter the optimal fiber length. More interesting is that the relative contribution of the Raman gain is much higher for relatively low losses. Comparing Figure 10a,b, it can be seen that at losses up to 2 dB/m, the output power with SRS is significantly higher than without it, while at losses $\geq$3 dB/m, the Raman contribution becomes relatively small. Therefore, for the experimental implementation of the proposed amplifier, one should aim to manufacture a high-quality multicore tellurite fiber.

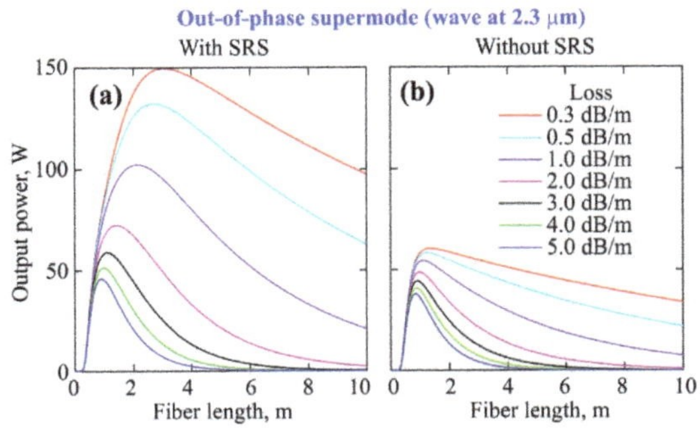

Figure 10. Powers vs. fiber length calculated with SRS (**a**) and without SRS (**b**) for waves at 2.3 μm propagating in out-of-phase supermode for pump power of 300 W and $N_{Tm} = 4 \times 10^{20}$ cm^{-3} and for different fiber losses. The legend in (**b**) is shared for both panels.

4. Discussion and Conclusions

To conclude, we have shown numerically that a trivalent-thulium-ion-doped TZ multi-core fiber with coupled axially symmetrical cores arranged in a symmetrical ring structure is a promising solution for diode-pumped optical amplifiers of 100-W power class at 2.3 μm, which is required for important applications. We studied the use of dual-wavelength seed signals with a power of order 1 mW at 1.96 μm and 2.3 μm and a commercially available diode pump at 793 nm with a power of up to 300 W. We demonstrated the possibility of attaining up to 50% conversion efficiency from the pump power to the wave at 2.3 μm which exceeds the Stokes limit by 15% due to a combination of laser amplification and SRS between signal waves (for losses of 0.3 dB/m). The proposed scheme can provide significantly higher efficiency in comparison to trivalent-thulium-ion-doped fiber lasers and amplifiers that do not benefit from additional Raman energy transfer [4–7]. Note that the use of a fiber based on TZ glass is substantial, since its Raman shift of 22.5 THz coincides with the difference in the frequencies for two different radiative transitions of the trivalent thulium ion. For other glasses that are transparent in this range, the Raman shift is smaller (~17 THz for ZBLAN [36], 10 THz for arsenic [37], ~7 THz for arsenic triselenide [37]), which is not suitable for implementing the concept of combining cascade laser amplification and SRS. We emphasize that the use of the out-of-phase supermode is important since its overlap integral with doped cores is higher compared to the commonly used in-phase supermode, whereas the out-of-phase supermode provides a higher total gain. In addition, it was previously shown that the out-of-phase supermode is stable at high powers, while the in-phase one is unstable [28–30]. As was shown in [30], small perturbations of the ideal transverse fiber structure led to a slight deformation of the out-of-phase mode, but do not lead to qualitative changes. Therefore, when deviating from the ideal geometry, only minor quantitative changes in the amplifier parameters are expected.

Note that the proposed design is especially good for high pump powers > 150 W. It provides high total power with an allowable load on each core. In this case, all powers refer to peak values. At lower target peak powers, a similar optimization can be performed using fewer cores, thereby reducing the effective mode area and increasing the Raman gain. One can also use a smaller cladding diameter to increase the overlap integral of the pump with doped cores. We considered CW waves, but the developed approach is valid for pulsed amplifiers with signal durations notably longer than the total lifetime of level $^3F_4$ (3 ms).

Author Contributions: Conceptualization, E.A.A.; methodology, E.A.A.; software, E.A.A.; validation, E.A.A. and A.V.A.; formal analysis, E.A.A. and A.V.A.; investigation, E.A.A.; data curation, E.A.A.; writing—original draft preparation, E.A.A.; writing—review and editing, A.V.A. and A.G.L.; visualization, E.A.A.; supervision, A.G.L.; funding acquisition, A.G.L. All authors have read and agreed to the published version of the manuscript.

Funding: This research was funded by the Russian Science Foundation, grant no. 23-12-00248 (https://rscf.ru/en/project/23-12-00248/ (accessed on 16 June 2023)).

Institutional Review Board Statement: Not applicable.

Informed Consent Statement: Not applicable.

Data Availability Statement: The data underlying the results presented in this article may be obtained from the authors upon reasonable request.

Conflicts of Interest: The authors declare no conflict of interest.

References

1. Moskalev, I.; Mirov, S.; Mirov, M.; Vasilyev, S.; Smolski, V.; Zakrevskiy, A.; Gapontsev, V. 140 W Cr:ZnSe Laser System. *Opt. Express* **2016**, *24*, 21090–21104. [CrossRef] [PubMed]
2. Zha, F.; Yu, X.; Chu, H.; Pan, H.; Zhao, S.; Loiko, P.; Pan, Z.; Li, D. Compact Diode-Pumped Continuous Wave and Passively Q Switched Tm:YAG Laser at 2.33 μm. *Opt. Lett.* **2022**, *47*, 6265–6268. [CrossRef] [PubMed]

3. Loiko, P.; Camy, P.; Soulard, R.; Guillemot, L.; Brasse, G.; Doualan, J.-L.; Braud, A.; Tyazhev, A.; Hideur, A.; Druon, F. Efficient Tm:LiYF4 Lasers at ~2.3 µm: Effect of Energy-Transfer Upconversion. *IEEE J. Quantum Electron.* **2019**, *55*, 1700212. [CrossRef]
4. Li, X.; Zhu, X.; Yang, L.; Cui, Y.; Zhou, Z.; Wang, M.; Wang, Z. High-Gain Single-Frequency Tm3+-Doped ZBLAN Fiber Amplifier at 2.33 µm. *Opt. Lett.* **2023**, *48*, 502–505. [CrossRef] [PubMed]
5. Muravyev, S.V.; Anashkina, E.A.; Andrianov, A.V.; Dorofeev, V.V.; Motorin, S.E.; Koptev, M.Y.; Kim, A.V. Dual-Band Tm^{3+}-Doped Tellurite Fiber Amplifier and Laser at 1.9 µm and 2.3 µm. *Sci. Rep.* **2018**, *8*, 16164. [CrossRef]
6. Denker, B.I.; Dorofeev, V.V.; Galagan, B.I.; Koltashev, V.V.; Motorin, S.E.; Plotnichenko, V.G.; Sverchkov, S.E. 2.3 µm Laser Action in Tm^{3+}-Doped Tellurite Glass Fiber. *Laser Phys. Lett.* **2018**, *16*, 015101. [CrossRef]
7. Denker, B.I.; Dorofeev, V.V.; Galagan, B.I.; Koltashev, V.V.; Motorin, S.E.; Plotnichenko, V.G.; Sverchkov, S.E. A 200 mW, 2.3 µm Tm3+-Doped Tellurite Glass Fiber Laser. *Laser Phys. Lett.* **2020**, *17*, 095101. [CrossRef]
8. Tao, G.; Ebendorff-Heidepriem, H.; Stolyarov, A.M.; Danto, S.; Badding, J.V.; Fink, Y.; Ballato, J.; Abouraddy, A.F. Infrared fibers. *Adv. Opt. Photonics* **2015**, *7*, 379–458. [CrossRef]
9. Bufetov, I.A.; Kosolapov, A.F.; Pryamikov, A.D.; Gladyshev, A.V.; Kolyadin, A.N.; Krylov, A.A.; Yatsenko, Y.P.; Biriukov, A.S. Revolver Hollow Core Optical Fibers. *Fibers* **2018**, *6*, 39. [CrossRef]
10. Dorofeev, V.V.; Moiseev, A.N.; Churbanov, M.F.; Kotereva, T.V.; Chilyasov, A.V.; Kraev, I.A.; Pimenov, V.G.; Ketkova, L.A.; Dianov, E.M.; Plotnichenko, V.G.; et al. Production and Properties of High Purity $TeO_2-WO_3-(La_2O_3, Bi_2O_3)$ and TeO2$-ZnO-Na_2O-Bi_2O_3$ Glasses. *J. Non-Cryst. Solids* **2011**, *357*, 2366–2370. [CrossRef]
11. Linganna, K.; In, J.-H.; Kim, S.H.; Han, K.; Choi, J.H. Engineering of TeO_2-ZnO-BaO-Based Glasses for Mid-Infrared Transmitting Optics. *Materials* **2020**, *13*, 5829. [CrossRef]
12. Strutynski, C.; Evrard, M.; Le Gendre, A.; Maldonado, A.; Désévédavy, F.; Gadret, G.; Jules, J.-C.; Smektala, F. Physicochemical Properties and Fiber-Drawing Ability of Tellurite Glasses in the TeO_2-ZnO-Y_2O_3 Ternary System. *Materials* **2022**, *15*, 1177. [CrossRef]
13. Gunha, J.V.; Muniz, R.F.; Somer, A.; Gonçalves, A.; Denkiewicz, R.; de Souza Antero, G.; Sales, T.O.; Jacinto, C.; El-Mallawany, R.; Novatski, A. Characterization of Oxyfluorotellurite Glasses with TeO_2–Li2O–ZnO–LiF Composition. *Ceram. Int.* **2022**, *48*, 4302–4311. [CrossRef]
14. O'Donnell, M.D.; Richardson, K.; Stolen, R.; Seddon, A.B.; Furniss, D.; Tikhomirov, V.K.; Rivero, C.; Ramme, M.; Stegeman, R.; Stegeman, G.; et al. Tellurite and Fluorotellurite Glasses for Fiber optic Raman Amplifiers: Glass Characterization, Optical Properties, Raman Gain, Preliminary Fiberization, and Fiber Characterization. *J. Am. Ceram. Soc.* **2007**, *90*, 1448–1457. [CrossRef]
15. Gomes, L.; Lousteau, J.; Milanese, D.; Scarpignato, G.C.; Jackson, S.D. Energy Transfer and Energy Level Decay Processes in Tm^{3+}-Doped Tellurite Glass. *J. Appl. Phys.* **2012**, *111*, 063105. [CrossRef]
16. Yao, T.; Huang, L.; Zhou, P.; Lei, B.; Leng, J.; Chen, J. Power Scaling on Tellurite Glass Raman Fibre Lasers for Mid-Infrared Applications. *High Pow Laser Sci. Eng.* **2018**, *6*, e24. [CrossRef]
17. Plotnichenko, V.G.; Sokolov, V.O.; Koltashev, V.V.; Dianov, E.M.; Grishin, I.A.; Churbanov, M.F. Raman Band Intensities of Tellurite Glasses. *Opt. Lett.* **2005**, *30*, 1156–1158. [CrossRef] [PubMed]
18. Jauregui, C.; Limpert, J.; Tünnermann, A. High-Power Fibre Lasers. *Nat. Photon.* **2013**, *7*, 861–867. [CrossRef]
19. Zervas, M.N.; Codemard, C.A. High Power Fiber Lasers: A Review. *IEEE J. Select. Topics Quantum Electron.* **2014**, *20*, 219–241. [CrossRef]
20. Cristiani, I.; Lacava, C.; Rademacher, G.; Puttnam, B.J.; Luìs, R.S.; Antonelli, C.; Mecozzi, A.; Shtaif, M.; Cozzolino, D.; Bacco, D.; et al. Roadmap on Multimode Photonics. *J. Opt.* **2022**, *24*, 083001. [CrossRef]
21. Lin, X.-J.; Gao, Y.-X.; Long, J.-G.; Wu, J.-W.; Li, X.-Y.; Hong, W.-Y.; Cui, H.; Luo, Z.-C.; Xu, W.-C.; Luo, A.-P. Spatial Beam Self-Cleaning in Bi-Tapered Multimode Fibers. *Photonics* **2021**, *8*, 479. [CrossRef]
22. Trikshev, A.I.; Kurkov, A.S.; Tsvetkov, V.B.; Filatova, S.A.; Kertulla, J.; Filippov, V.; Chamorovskiy, Y.K.; Okhotnikov, O.G. A 160 W Single-Frequency Laser Based on an Active Tapered Double-Clad Fiber Amplifier. *Laser Phys. Lett.* **2013**, *10*, 065101. [CrossRef]
23. Khudyakov, M.M.; Levchenko, A.E.; Velmiskin, V.V.; Bobkov, K.K.; Aleshkina, S.S.; Zaushitsyna, T.S.; Bubnov, M.M.; Yashkov, M.V.; Guryanov, A.N.; Kotov, L.V.; et al. Narrow-Linewidth Diffraction-Limited Tapered Er-Doped Fiber Amplifier with 2 mJ Pulse Energy. *Photonics* **2022**, *9*, 933. [CrossRef]
24. Klenke, A.; Jauregui, C.; Steinkopff, A.; Aleshire, C.; Limpert, J. High-Power Multicore Fiber Laser Systems. *Prog. Quantum Electron.* **2022**, *84*, 100412. [CrossRef]
25. Ortiz, A.M.; Sáez, R.L. Multi-Core Optical Fibers: Theory, Applications and Opportunities. *Sel. Top. Opt. Fiber Technol. Appl.* **2018**, *2*, 63–102. [CrossRef]
26. Wolf, A.; Dostovalov, A.; Bronnikov, K.; Skvortsov, M.; Wabnitz, S.; Babin, S. Advances in Femtosecond Laser Direct Writing of Fiber Bragg Gratings in Multicore Fibers: Technology, Sensor and Laser Applications. *Opto-Electron. Adv.* **2022**, *5*, 210055. [CrossRef]
27. Agrawal, G.P. *Nonlinear Fiber Optics*, 6th ed.; Elsevier: Amsterdam, The Netherlands, 2019.
28. Balakin, A.A.; Skobelev, S.A.; Anashkina, E.A.; Andrianov, A.V.; Litvak, A.G. Coherent Propagation of Laser Beams in a Small-Sized System of Weakly Coupled Optical Light-Guides. *Phys. Rev. A* **2018**, *98*, 043857. [CrossRef]
29. Balakin, A.A.; Skobelev, S.A.; Andrianov, A.V.; Anashkina, E.A.; Litvak, A.G. Coherent Amplification of High-Power Laser Radiation in Multicore Fibers from a Rectangular Array of Cores. *Opt. Lett.* **2021**, *46*, 246–249. [CrossRef]

30. Andrianov, A.V.; Kalinin, N.A.; Anashkina, E.A.; Egorova, O.N.; Lipatov, D.S.; Kim, A.V.; Semjonov, S.L.; Litvak, A.G. Selective Excitation and Amplification of Peak-Power-Scalable Out-of-Phase Supermode in Yb-Doped Multicore Fiber. *J. Light. Technol.* **2020**, *38*, 2464–2470. [CrossRef]
31. Tünnermann, H.; Shirakawa, A. Self-Focusing in Multicore Fibers. *Opt. Express* **2015**, *23*, 2436–2445. [CrossRef]
32. Bookey, H.T.; Lousteau, J.; Jha, A.; Gayraud, N.; Thomson, R.R.; Psaila, N.D.; Li, H.; MacPherson, W.N.; Barton, J.S.; Kar, A.K. Multiple Rare Earth Emissions in a Multicore Tellurite Fiber with a Single Pump Wavelength. *Opt. Express* **2007**, *15*, 17554–17561. [CrossRef] [PubMed]
33. Cheng, T.; Duan, Z.; Gao, W.; Asano, K.; Liao, M.; Deng, D.; Suzuki, T.; Ohishi, Y. A Novel Seven-Core Multicore Tellurite Fiber. *J. Light. Technol.* **2013**, *31*, 1793–1796. [CrossRef]
34. Anashkina, E.A.; Andrianov, A.V. Design and Dispersion Control of Microstructured Multicore Tellurite Glass Fibers with In-Phase and Out-of-Phase Supermodes. *Photonics* **2021**, *8*, 113. [CrossRef]
35. Kalinin, N.A.; Anashkina, E.A.; Egorova, O.N.; Zhuravlev, S.G.; Semjonov, S.L.; Kim, A.V.; Litvak, A.G.; Andrianov, A.V. Controlled Excitation of Supermodes in a Multicore Fiber with a 5 × 5 Square Array of Strongly Coupled Cores. *Photonics* **2021**, *8*, 314. [CrossRef]
36. Yan, X.; Kito, C.; Miyoshi, S.; Liao, M.; Suzuki, T.; Ohishi, Y. Raman Transient Response and Enhanced Soliton Self-Frequency Shift in ZBLAN Fiber. *J. Opt. Soc. Am. B* **2012**, *29*, 238–243. [CrossRef]
37. Li, W.; Seal, S.; Rivero, C.; Lopez, C.; Richardson, K.; Pope, A.; Schulte, A.; Myneni, S.; Jain, H.; Antoine, K.; et al. Role of S/Se Ratio in Chemical Bonding of As–S–Se Glasses Investigated by Raman, x-Ray Photoelectron, and Extended x-Ray Absorption Fine Structure Spectroscopies. *J. Appl. Phys.* **2005**, *98*, 053503. [CrossRef]

photonics

MDPI

Communication

Distributed Bragg Reflector Laser Based on Composite Fiber Heavily Doped with Erbium Ions

Mikhail I. Skvortsov [1,*], Kseniya V. Proskurina [1], Evgeniy V. Golikov [1], Alexander V. Dostovalov [1], Vadim S. Terentyev [1], Olga N. Egorova [2], Sergey L. Semjonov [3] and Sergey A. Babin [1]

1 Institute of Automation and Electrometry, Siberian Branch of the Russian Academy of Sciences, 1 Ac. Koptyug Ave., 630090 Novosibirsk, Russia; ksyna_98@mail.ru (K.V.P.); golikov.inc@mail.ru (E.V.G.); dostovalov@iae.nsk.su (A.V.D.); terentyev@iae.nsk.su (V.S.T.); babin@iae.nsk.su (S.A.B.)

2 Prokhorov General Physics Institute of the Russian Academy of Science, 38 Vavilov St., 119991 Moscow, Russia; egorova@nsc.gpi.ru

3 Dianov Fiber Optic Research Center, Prokhorov General Physics Institute of the Russian Academy of Science, 38 Vavilov St., 119991 Moscow, Russia; sls@fo.gpi.ru

* Correspondence: skvorczov@iae.nsk.su

Abstract: A distributed Bragg reflector (DBR) laser with a specially designed, heavily Er^{3+}-doped composite fiber of a length as short as 1.8 cm is demonstrated. The DBR laser, pumped by a 980 nm laser diode with power of up to 370 mW, generates single-frequency radiation at a wavelength of 1535 nm with a narrow instantaneous linewidth of <100 Hz and a high output power of 2 mW. The obtained Er^{3+}-doped fiber laser parameters pave the way toward a broad range of practical applications from telecommunications and sensing to scientific research.

Keywords: fiber laser; fiber Bragg gratings; distributed Bragg reflectors; composite erbium fiber; single-frequency regime; narrow linewidth

Citation: Skvortsov, M.I.; Proskurina, K.V.; Golikov, E.V.; Dostovalov, A.V.; Terentyev, V.S.; Egorova, O.N.; Semjonov, S.L.; Babin, S.A. Distributed Bragg Reflector Laser Based on Composite Fiber Heavily Doped with Erbium Ions. *Photonics* **2023**, *10*, 679. https://doi.org/10.3390/photonics10060679

Received: 11 May 2023
Revised: 8 June 2023
Accepted: 9 June 2023
Published: 12 June 2023

1. Introduction

Single-longitudinal-mode (SLM) fiber lasers are widely used in a broad range of applications such as coherent telecommunications [1], sensing with a high spatial resolution [2,3], Rayleigh backscattering reflectometry [4] and high-resolution spectroscopy [2]. Distributed feedback (DFB) fiber lasers are the most popular among them and provide stable single-frequency generation due to the spectral filtering in a laser cavity formed by a fiber Bragg grating with a pi-phase shift inscribed in the core of the active fiber [5–7]. The fabrication of a phase-shifted FBG with an overall length of ≥5 cm is quite a challenging task in comparison with uniform FBG inscription, considering the precise spatial position and amplitude of the phase shift to be created during the inscription process. For these reasons, elaborated phase masks with phase shifts in the structure or complicated point-by-point inscription techniques are used for the inscription process via continuous UV radiation [7] and femtosecond (fs) laser pulses [8], respectively. Aside from DFB fiber lasers, it is possible to achieve single-frequency generation with comparable output parameters using a distributed Bragg reflector (DBR) laser scheme characterized by a short cavity with an active fiber between two long, uniform FBGs [9]. For example, an SLM fiber laser based on active an Er^{3+}/Yb^{3+} co-doped fiber was demonstrated, emitting at 1560 nm with an output power of up to 200 mW and a spectral width of 2 kHz [10]. A reduction in the cavity length of a DBR laser based on an Er^{3+}/Yb^{3+} active fiber to 2 cm in length was achieved in [11], in which one FBG was replaced by a dielectric mirror, allowing for laser generation at 1535 nm with an output power of up to 300 mW and a spectral width of 1.6 kHz.

However, the development of a DBR fiber laser with single-frequency generation based on an Er^{3+} active fiber is difficult task owing to the low absorption cross-sections of Er^{3+} active ions and a clustering problem in the case of a high concentration of Er^{3+}

ions, which results in a pulsed regime of laser generation. To avoid these disadvantages, a specially designed Er^{3+}-doped composite was developed [12]. The measured gains at the wavelength of 1535 nm were about 1.6 dB/cm and 3.1 dB/cm for fibers with the erbium oxide concentrations of 1 wt% and 3 wt%, respectively [13]. Owing to a high radiation absorption in the case of the 3 wt% active fiber, an extremely short DFB fiber laser with a 5 mm long cavity was demonstrated through the use of a point-by-point femtosecond laser inscription technique of a pi-phase shifted FBG [14]. Moreover, using a 1 wt Er^{3+} active fiber in a DFB laser configuration with a cavity length of 5 cm, single-frequency generation lasing at 1559.5 nm with a output power up to 3 mW at a pump power of 320 mW was demonstrated [15].

In this work, we present the results of the development of a DBR laser with an Er^{3+} active fiber with a high concentration of active ions (3 wt%) in a cavity as short as 2 cm formed by two mirrors: a less reflective FBG inscribed in the active fiber core and a highly reflective dielectric multilayer mirror deposited on the fiber end face. A single-frequency lasing regime is observed in the whole pump power range (up to 370 mW) with the instantaneous laser linewidth of ~100 Hz at a maximum output power of 2.05 mW. To the best of our knowledge, the DBR Er^{3+}-doped fiber laser developed here has a record short cavity length of 1.8 cm, resulting in spectral and output power parameters favorable for a wide range of applications.

2. Experiment

The optical fiber used as an active medium for the DFB lasers was produced via a rod-in-tube technique, using a phosphate glass rod and a silica tube and then drawing the preform. The process of manufacturing the fiber is described in detail in papers [16,17]. To fabricate the core, we used glass of the same composition as the glass used in [16,17]. In addition to 65 mol% of phosphorus oxide (PO_2), this composition contained 7 mol% of Al_2O_3, 12 mol% of B_2O_3, 9 mol% of Li_2O, and 7 mol% of RE_2O_3. The concentration of erbium oxide in the initial glass was about 1.2 mol% (3 wt% erbium). The core diameter of the active fiber amounted to 3.6 μm, which corresponded to a mode field diameter of 4.34 μm. The peak absorptions were about 1.25 dB/cm and 3.65 dB/cm at a pump wavelength of 980 nm and an output radiation wavelength of 1535 nm, respectively. Thus, due to the high concentration of active ions, $N \approx 1.6 \times 10^{20}$ cm^{-3}, the fiber gain coefficient was as high as $g \approx 3.1$ dB/cm at a lasing wavelength of 1535 nm. At such a high concentration of active ions, ion clustering usually occurs, leading to pulsed laser generation [18,19]. A high phosphorus concentration in the core of the composite fiber reduced the clustering effect so that the overall concentration of clusters decreased to ~8% [14], which is significantly lower than cluster concentration in commercial active fibers with lower concentrations of ions. For example, the cluster concentration in the Er-doped active fiber Er-80 amounts to 14% at a lower ion concentration ($N = 3.7 \times 10^{19}$ cm^{-3}).

The highly reflective dielectric mirror used in the DBR laser cavity was deposited via magnetron sputtering on the active fiber end face [20]. The mirror consisted of 15 quarter-wavelength layers of $H(LH)_7$ with a high refractive index, $nTiO_2 \approx 2.4$ (H, titanium oxide TiO_2), and a low refractive index, $nSiO_2 \approx 1.47$ (L, SiO_2), resulting in reflection coefficient of >99% at 1.55 μm and a total thickness of ≈ 3.2 μm. The reflection coefficient at a pump wavelength of 980 nm amounted to 20%. An FBG with a narrow spectral bandwidth inscribed via a femtosecond laser point-by-point technique [8] was used as an output coupler with a reflection wavelength of ≈1535 nm, a reflection coefficient of 98%, a spectral bandwidth of ≈60 pm and a length of 1.5 cm. The required length of the active fiber for CW laser generation depends on the gain coefficient at the output radiation wavelength and the total losses in the cavity:

$$gL = T_{FBG} + T_{dm} + \alpha_L + \ln(1 - \alpha_{splice}), \quad (1)$$

where L is the length of the active fiber, and g is the gain coefficient, which was estimated to be 3.1 dB/cm at a wavelength of 1535 nm. T_{FBG} and T_{dm} are the losses of the FBG and dielectric mirror deposited on the fiber tip, respectively. The value of the passive loss α in the fiber, measured at a wavelength of 1300 nm, was about 4–5 dB/m [13]. Due to the difference in the mode field diameter (MFD) between the SMF-28 and the composite fiber, the splicing losses α_{splice} amounted to $\approx$50%. Despite the high total losses, the minimum fiber length required for the CW laser generation was estimated to be $\approx$1 cm. In the experiment, the length of the active fiber was 1.8 cm, taking into account that an FBG with a length of 1.5 cm results in a total resonator length of 3.3 cm.

The experimental scheme of the DBR laser is shown in Figure 1. The DBR fiber cavity was pumped by a single-mode laser diode with a wavelength of 980 nm and an output power of up to 370 mW through the 980/1.550 nm wavelength division multiplexer (WDM). The laser wavelength and output power were measured using a Yokogawa AQ6370 Optical Spectrum Analyzer (OSA) with a resolution of 20 pm. An Agilent N9010A Radio Frequency (RF) spectrum analyzer and a Thorlabs DET08CFC 5 GHz photodiode were used to measure the relative intensity noise (RIN). A Mach-Zehnder interferometer (MZI) was utilized to determine both the instantaneous laser linewidth and the spectral width at longer time intervals of ~100 µsec. One of the MZI arms contained a 25 km fiber, and the other arm contained an acousto-optic modulator (AOM) controlled by an Agilent 33250A signal generator with a carrier frequency of 80 MHz. The beat signal was measured with a Thorlabs DET08CFC photodiode and a LeCroy WavePro 725Zi-A/5 GHz oscilloscope with FFT function, which allowed for the waveforms and RF spectra of the signal to be measured.

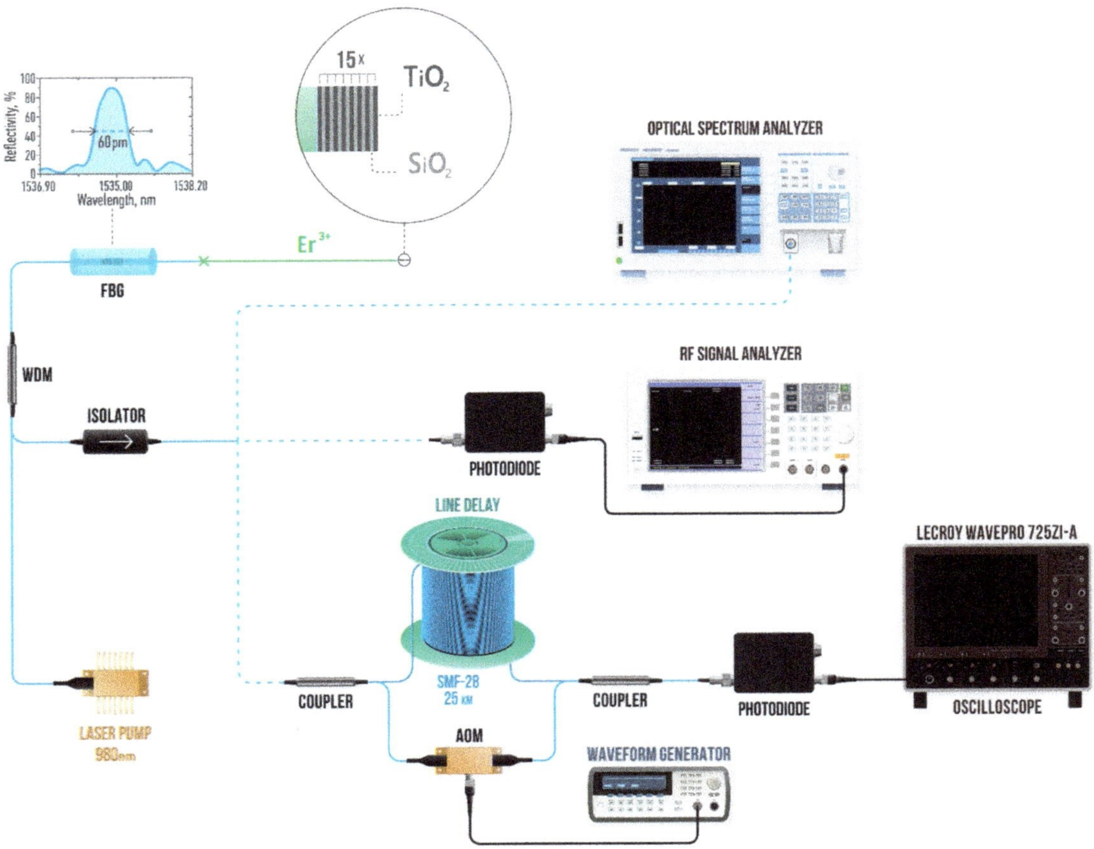

Figure 1. Experimental setup.

3. Results and Discussion

Figure 2a shows the measured output power as a function of the pump power. The threshold for generating radiation with a wavelength of ≈1535 nm was reached at a pump power of 140 mW. The high value of the threshold power can be explained by the losses of splicing between the active fiber and the single-mode fiber with the FBG. The output power reached 2.05 mW at a maximum pump power of 370 mW, which corresponds to a differential efficiency of η ~0.85% and significantly exceeds this value for typical Er^{3+} DFB lasers [21].

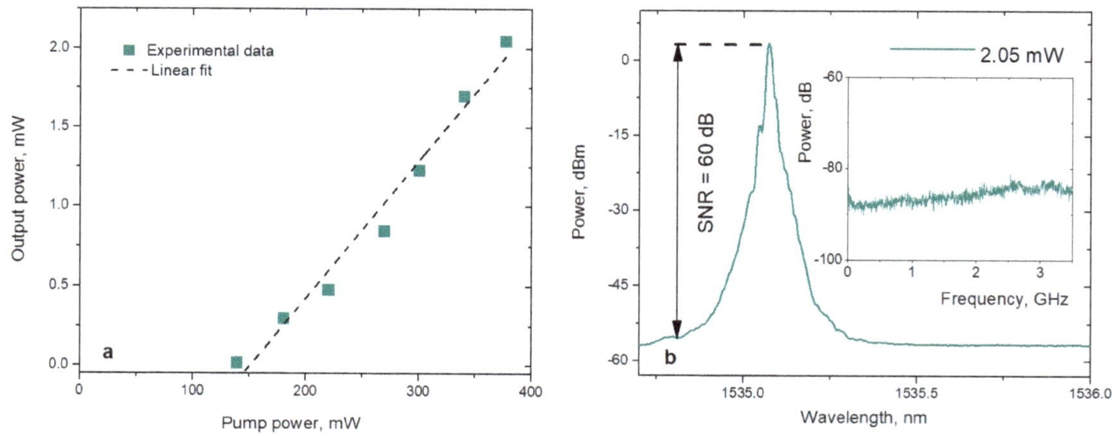

Figure 2. (**a**) The measured output power as a function of pump power; (**b**) output laser radiation spectrum at maximum output power; inset: RF spectrum of radiation.

Figure 2b shows the optical spectrum of the output radiation at maximum output power: the spectral width at the half-maximum of 20 pm corresponds to the OSA instrumental function, and the signal-to-noise ratio is 60 dB. The RF spectrum at the maximum output power presented in the inset of Figure 2b reveals no longitudinal mode beat peaks in the frequency range up of to 3.6 GHz (≈28 pm at 1535 nm), thus confirming the regime of single-frequency generation.

The measured level of the RIN was −90 dB/Hz at 527 kHz, which is a typical value for this type of laser without external active stabilization. Using the technique of beat waveform processing described in [22–24], the frequency noise spectrum was obtained (see Figure 3b). The value of the instantaneous laser linewidth was determined by the level of white noise, S_0, and was $\Delta v = \pi S_0 \approx 94$ Hz [25]. The relatively high value of the instantaneous linewidth was determined by the high power level of the spontaneous noise [26], which depends on the concentration of active ions in the fiber [27]. The self-heterodyne technique was used to determine the laser linewidth at the delay line time interval [28]. Figure 3c shows an RF beat signal with a width of 110 kHz at a level of −20 dB, which corresponds to the laser linewidth of 5.5 kHz for the Lorenz profile. Therefore, the obtained spectral characteristics of the proposed DBR-laser are comparable with the spectral characteristics of typical Er^{3+} DFB lasers but with a significant excess of output power.

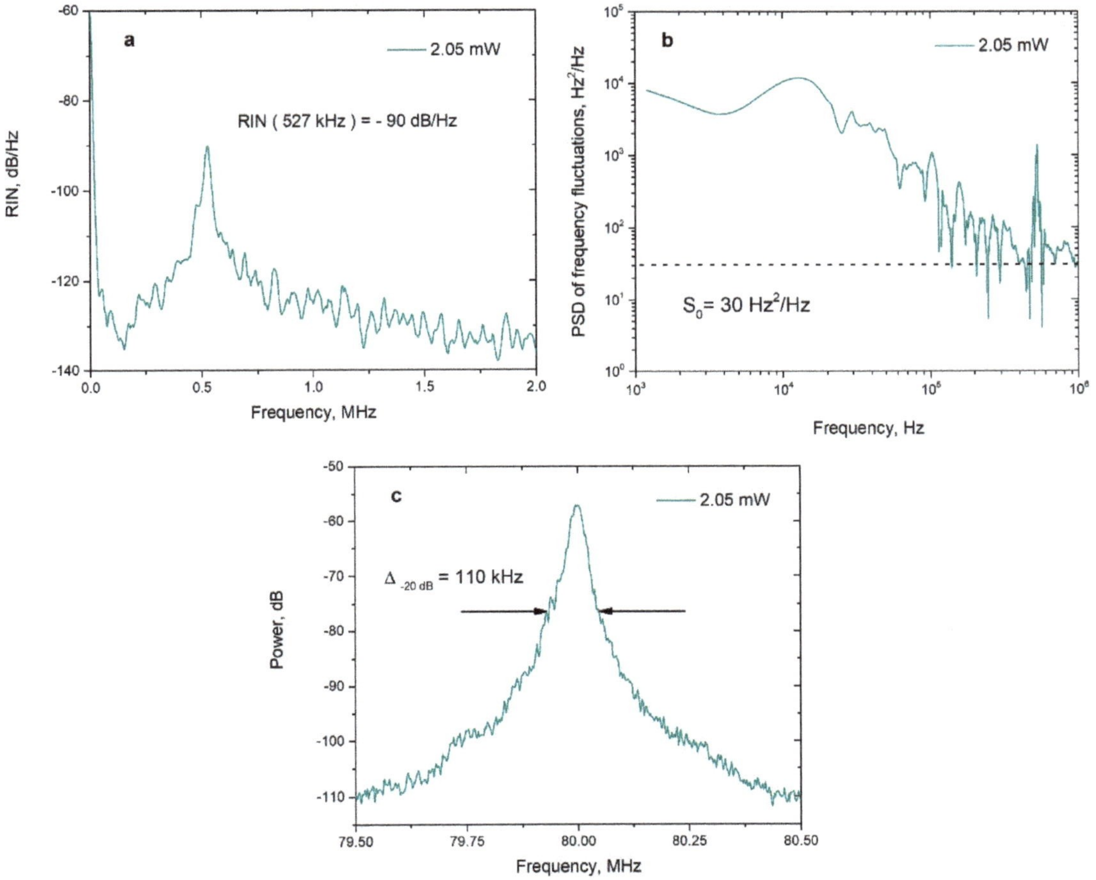

Figure 3. (**a**) Relative intensity noise measured at the maximum output power of the laser; (**b**) PSD of frequency fluctuations; (**c**) RF beating spectra obtained via heterodyne.

4. Conclusions

Thus, based on a heavily Er-doped fiber with a gain coefficient of ≈3.1 dB/cm, a DBR fiber laser has been developed with a record short active fiber length of 1.8 cm. A significant shortening of the cavity length was achieved through the use of a dielectric, highly reflective thin-film mirror deposited on the end face of the active fiber. In combination with the FBG, the total length of the structure was 3.3 cm. Due to the broad free spectral range of the obtained short laser cavity, a single-frequency regime was observed in the entire lasing power range with a maximum output power of 2 mW. The signal-to-noise ratio was at least 60 dB, as measured by the OSA, with an optical bandwidth resolution of 20 pm, and the peak of the RIN was observed at a frequency of 527 kHz; the peak amplitude value corresponded to −90 dB/Hz. The measured instantaneous linewidth of the laser at maximum output power was 100 Hz, and the width measured over a time interval of ~100 μs was 5.5 kHz. In addition, the use of a broadband dielectric mirror makes it possible to easily tune the generation wavelength via the compression/tension of the output FBG.

The characteristics of the developed compact DBR fiber laser presented herein open the path for applications in telecommunications, sensing, metrology, and scientific research in which compact and stable narrowband laser sources with tuning options are required [2].

Author Contributions: Conceptualization, S.A.B., S.L.S. and M.I.S.; data curation, S.A.B., M.I.S., V.S.T. and K.V.P.; investigation, K.V.P., M.I.S., E.V.G., A.V.D. and V.S.T.; writing—review and editing, M.I.S., O.N.E., A.V.D., V.S.T. and S.A.B.; funding acquisition, S.A.B., S.L.S. and O.N.E.; formal analysis, E.V.G.; writing—original draft preparation, K.V.P.; project administration, S.A.B. and S.L.S.; supervision, O.N.E. All authors have read and agreed to the published version of the manuscript.

Funding: The study is supported by the state budget of IA&E SB RAS (project No. 121030500067-5). Experimental studies were carried out using the equipment of the Center for Collective Use "Spectroscopy and Optics" at the Institute of Automation and Electrometry, SB RAS.

Institutional Review Board Statement: Not applicable.

Informed Consent Statement: Not applicable.

Data Availability Statement: Not applicable.

Conflicts of Interest: The authors declare no conflict of interest.

References

1. Beppu, S.; Kasai, K.; Yoshida, M.; Nakazawa, M. 2048 QAM (66 Gbit/s) Single-Carrier Coherent Optical Transmission over 150 km with a Potential SE of 15.3 Bit/s/Hz. *Opt. Express* **2015**, *23*, 4960–4969. [CrossRef] [PubMed]
2. Persijn, S.; Harren, F.; Van Der Veen, A. Quantitative Gas Measurements Using a Versatile OPO-Based Cavity Ringdown Spectrometer and the Comparison with Spectroscopic Databases. *Appl. Phys. B Lasers Opt.* **2010**, *100*, 383–390. [CrossRef]
3. Cranch, G.A.; Flockhart, G.M.H.; Kirkendall, C.K. Distributed Feedback Fiber Laser Strain Sensors. *IEEE Sens. J.* **2008**, *8*, 1161–1172. [CrossRef]
4. Shatalin, S.V.; Treschikov, V.N.; Rogers, A.J. Interferometric Optical Time-Domain Reflectometry for Distributed Optical-Fiber Sensing. *Appl. Opt.* **1998**, *37*, 5600–5604. [CrossRef] [PubMed]
5. Loh, W.H.; Laming, R.I. 1.55 Mkm Phase-Shifted Distributed Feedback Fibre Laser. *Electron. Lett.* **1995**, *31*, 1440–1442. [CrossRef]
6. Butov, O.V.; Rybaltovsky, A.A.; Vyatkin, M.Y.; Bazakutsa, A.P.; Popov, S.M.; Chamorovskiy, Y.K.; Golant, K.M. 1030 nm Yb^{3+} Distributed Feedback Short Cavity Silica-Based Fiber Laser. *J. Opt. Soc. Am. B* **2017**, *34*, 43–48. [CrossRef]
7. Babin, S.A.; Churkin, D.V.; Ismagulov, A.E.; Kablukov, S.I.; Nikulin, M.A. Single Frequency Single Polarization DFB Fiber Laser. *Laser Phys. Lett.* **2007**, *4*, 428–432. [CrossRef]
8. Dostovalov, A.V.; Wolf, A.A.; Parygin, A.V.; Zyubin, V.E.; Babin, S.A. Femtosecond Point-by-Point Inscription of Bragg Gratings by Drawing a Coated Fiber through Ferrule. *Opt. Express* **2016**, *24*, 16232–16237. [CrossRef]
9. Kelson, I.; Hardy, A.A. Strongly Pumped Fiber Lasers. *IEEE J. Quantum Electron.* **1998**, *34*, 1570–1577. [CrossRef]
10. Spiegelberg, C.; Geng, J.; Hu, Y.; Kaneda, Y.; Jiang, S.; Peyghambarian, N. Low-Noise Narrow-Linewidth Fiber Laser at 1550 nm. *J. Light. Technol.* **2004**, *22*, 57–62. [CrossRef]
11. Xu, S.H.; Yang, Z.M.; Liu, T.; Zhang, W.N.; Feng, Z.M.; Zhang, Q.Y.; Jiang, Z.H. An Efficient Compact 300 mW Narrow-Linewidth Single Frequency Fiber Laser at 1.5 μm. *Opt. Express* **2010**, *18*, 1249–1254. [CrossRef]
12. Rybaltovsky, A.A.; Egorova, O.N.; Zhuravlev, S.G.; Galagan, B.I.; Sverchkov, S.E.; Denker, B.I.; Semjonov, S.L. Distributed Bragg Reflector Fiber Laser Directly Written in a Composite Fiber Manufactured by Melting Phosphate Glass in a Silica Tube. *Opt. Lett.* **2019**, *44*, 3518–3521. [CrossRef]
13. Denker, B.I.; Galagan, B.I.; Kamynin, V.A.; Ponosova, A.A.; Riumkin, K.E.; Semjonov, S.L.; Sverchkov, S.E.; Tsvetkov, V.B. Gain Characteristics of Fibers with a Heavily Erbium-Doped Phosphate-Based Core and Silica Cladding. *J. Opt. Soc. Am. B* **2019**, *36*, 2705. [CrossRef]
14. Skvortsov, M.I.; Wolf, A.A.; Vlasov, A.A.; Proskurina, K.V.; Dostovalov, A.V.; Egorova, O.N.; Galagan, B.I.; Sverchkov, S.E.; Denker, B.I.; Semjonov, S.L.; et al. Advanced Distributed Feedback Lasers Based on Composite Fiber Heavily Doped with Erbium Ions. *Sci. Rep.* **2020**, *10*, 14487. [CrossRef]
15. Egorova, O.N.; Medvedkov, O.I.; Seregin, E.S.; Vasil, S.A.; Sverchkov, S.E.; Galagan, B.I.; Denker, B.I.; Danielyan, G.L.; Pustovoi, V.I.; Semjonov, S.L. Single-Frequency Fibre Laser with a Cavity Formed by Bragg Gratings Written in the Core of an Active Composite Fibre Using KrF Laser Radiation Single-Frequency Fibre (248 nm). *Quantum Electron.* **2019**, *49*, 2–7. [CrossRef]
16. Egorova, O.N.; Semjonov, S.L.; Velmiskin, V.V.; Yatsenko, Y.P.; Sverchkov, S.E.; Galagan, B.I.; Denker, B.I.; Dianov, E.M. Phosphate-Core Silica-Clad Er/Yb-Doped Optical Fiber and Cladding Pumped Laser. *Opt. Express* **2014**, *22*, 7632. [CrossRef]
17. Egorova, O.N.; Semjonov, S.L.; Medvedkov, O.I.; Astapovich, M.S.; Okhrimchuk, A.G.; Galagan, B.I.; Denker, B.I.; Sverchkov, S.E.; Dianov, E.M. High-Beam Quality, High-Efficiency Laser Based on Fiber with Heavily Yb^{3+}-Doped Phosphate Core and Silica Cladding. *Opt. Lett.* **2015**, *40*, 3762. [CrossRef]
18. Sanchez, F.; Le Boudec, P.; François, P.-L.; Stephan, G. Effects of Ion Pairs on the Dynamics of Erbium-Doped Fiber Lasers. *Phys. Rev. A* **1993**, *48*, 2220–2229. [CrossRef]
19. Li, N.; Bradley, J.D.; Singh, G.; Magden, E.S.; Sun, J.; Watts, M.R. Self-Pulsing in Erbium-Doped Fiber Laser. *Optoelectron. Glob. Conf.* **2015**, *16*, 15–16.

20. Born, M.; Wolf, E. *Principles of Optics: Electromagnetic Theory of Propagation, Interference and Diffraction of Light*; Elsevier: Amsterdam, The Netherlands, 2013; ISBN 148310320X.
21. Skvortsov, M.I.; Wolf, A.A.; Dostovalov, A.V.; Vlasov, A.A.; Akulov, V.A.; Babin, S.A. Distributed Feedback Fiber Laser Based on a Fiber Bragg Grating Inscribed Using the Femtosecond Point-by-Point Technique. *Laser Phys. Lett.* **2018**, *15*, 35103. [CrossRef]
22. Andel, N.; Yara, M.I.; Ohamed, M.; Ellahi, S.; Ouici, T.A.S.; Ardaillon, R.; Hilippe, P.; Ignoret, S. Time-Dependent Laser Linewidth: Beat-Note Digital Acquisition and Numerical Analysis. *Optics Express* **2016**, *24*, 401–407.
23. Nikitin, S.; Fomiryakov, E.; Kharasov, D.; Nanii, O.; Treshchikov, V. Characterization of Ultra-Narrow Linewidth Lasers for Phase-Sensitive Coherent Reflectometry Using EOM Facilitated Heterodyning. *J. Light. Technol.* **2020**, *38*, 1446–1453. [CrossRef]
24. Fomiryakov, E.; Kharasov, D.; Nikitin, S.; Nanii, O.; Treshchikov, V. New Approach to Laser Characterization Using Delayed Self-Heterodyne Interferometry. *J. Light. Technol.* **2021**, *39*, 5191–5196. [CrossRef]
25. Svelto, O. *Principles of Lasers*, 5th ed.; Springer: New York, NY, USA; Dordrecht, The Netherlands; Berlin/Heidelberg, Germany; London, UK, 2018; ISBN 9788578110796.
26. Skvortsov, M.I.; Abdullina, S.R.; Podivilov, E.V.; Vlasov, A.A.; Kharasov, D.R.; Fomiryakov, E.A.; Nikitin, S.P.; Treshchikov, V.N.; Babin, S.A. Extreme Narrowing of the Distributed Feedback Fiber Laser Linewidth Due to the Rayleigh Backscattering in a Single-Mode Fiber: Model and Experimental Test. *Photonics* **2022**, *9*, 590. [CrossRef]
27. Lei, C.; Feng, H.; Messaddeq, Y.; LaRochelle, S. Investigation of C-Band Pumping for Extended L-Band EDFAs. *J. Opt. Soc. Am. B* **2020**, *37*, 2345–2352. [CrossRef]
28. Okoshi, T.; Kikuchi, K.; Nakayama, A. Novel Method for High Resolution Measurement of Laser Output Spectrum. *Electron. Lett.* **1980**, *16*, 630–631. [CrossRef]

photonics

MDPI

Article

High Efficient Random Laser with Cavity Based on the Erbium-Doped Germanophosphosilicate Artificial Rayleigh Fiber

Sergei Popov [1,*], Andrey Rybaltovsky [2], Alexei Bazakutsa [3], Alexander Smirnov [3], Dmitry Ryakhovskiy [1], Viktor Voloshin [1], Alexander Kolosovskii [1], Igor Vorob'ev [1], Viktor Isaev [1], Yuriy Chamorovskiy [1], Denis Lipatov [4] and Oleg Butov [3]

1 Kotelnikov Institute of Radioengineering and Electronics (Fryazino Branch), Russian Academy of Sciences, Vvedensky Sq. 1, 141190 Fryazino, Moscow Region, Russia; dryh97@mail.ru (D.R.); v.voloshin@bk.ru (V.V.); kolos_ao@mail.ru (A.K.); infosiv@fryazino.net (I.V.); isaev@ms.ire.rssi.ru (V.I.); yurichamor@fireras.su (Y.C.)
2 Dianov Fiber Optics Research Center, Prokhorov General Physics Institute, Russian Academy of Sciences, St. Vavilova 38, 119333 Moscow, Russia; rybaltovsky@yandex.ru
3 Kotelnikov Institute of Radioengineering and Electronics, Russian Academy of Sciences, St. Mokhovaya 11-7, 125009 Moscow, Russia; a.bazakutsa@optel.ru (A.B.); alsmir1988@mail.ru (A.S.); obutov@mail.ru (O.B.)
4 G. G. Devyatykh Institute of Chemistry of High-Purity Substances, Russian Academy of Sciences, St. Tropinina 49, 603951 Nizhny Novgorod, Russia; lidenis@yandex.ru
* Correspondence: sergei@popov.eu.org; Tel.: +7-925-338-5702

Abstract: The Erbium "random" laser, based on the artificial Rayleigh fiber, has been comparatively studied in detail under two different pump conditions: 974.5 and 1485 nm pumping wavelengths. The artificial Rayleigh 7-m-long fiber was used as a laser cavity, it was formed by the ultraviolet (UV) inscription of the uniform array of the weakly reflective fiber Bragg grating (FBG) during the fiber drawing process. The UV photosensitivity of the Erbium-doped fiber originated from the specially developed (germanophosphosilicate) core glass composition. The emission spectrum of the fabricated "random" fiber laser had a single narrow peak at the 1548 nm wavelength. It was clearly revealed that the extension of the laser cavity by the separate wavelength-matched 90%-reflective FBG resulted in a significant laser efficiency growth. The highest laser slope efficiency of 33% and the laser output power of 80 mW were reached in the FBG-modified cavity at the 974.5-nm-wavelength pumping. The continuous-wave operation mode of this laser has been confirmed. The laser linewidth value measured by the delayed self-heterodyne technique was about 550 Hz.

Keywords: random fiber laser; fiber Bragg gratings array; artificial Rayleigh fiber; erbium-doped germanophosphosilicate optical fiber

Citation: Popov, S.; Rybaltovsky, A.; Bazakutsa, A.; Smirnov, A.; Ryakhovskiy, D.; Voloshin, V.; Kolosovskii, A.; Vorob'ev, I.; Isaev, V.; Chamorovskiy, Y.; et al. High Efficient Random Laser with Cavity Based on the Erbium-Doped Germanophosphosilicate Artificial Rayleigh Fiber. *Photonics* **2023**, *10*, 748. https://doi.org/10.3390/photonics10070748

Received: 31 May 2023
Revised: 26 June 2023
Accepted: 26 June 2023
Published: 28 June 2023

1. Introduction

Optical fibers (OF) are currently used for telecommunications, sensor systems and lasers. Currently, a new direction that is known as "random" fiber lasers [1–4] is actively developing. This direction of photonics has become a subject of great interest for researchers around the world due to the fact that random fiber lasers are able to generate light with unique performance characteristics without imposing strict requirements on the optical cavity. In this case, amplification is achieved due to the Raman scattering effects [2] or stimulated Brillouin scattering (SBS) [3]. The feedback in optical fibers is achieved due to the weak stationary ("frozen into the glass grid") scattering centers, uniformly distributed over a fiber length (Rayleigh scattering). This leads to the fact that the cavity of random lasers is constructed using long (1–100 km) OFs. The current trends in "random" fiber lasers are associated with the transition to lasers with cavities, based on an array of the weakly reflective, identical fiber Bragg gratings (FBGs) [5,6]. FBG arrays are also called "artificial Rayleigh fibers", as they are mainly used to increase the intensity of the return optical signal and reduce the cavity length because the reflection signal from Rayleigh scattering is extremely small [4] (−82 dB for pulses with duration of 1 ns). For random laser development, it is important to explore and fabricate the artificial Rayleigh fibers with optimal characteristics.

The process of forming such structures by UV radiation is carried out in several iterations: first, a polymer coating is removed from the OF section with a length of ~10 mm, then an FBG is formed in it using laser UV radiation, after which the OF section with the inscribed FBG is overcoated by polymer, and the whole procedure is repeated in the next section of the OF. The usage of the described technique contributes to a significant increase in the return signal, but also causes a deterioration in the mechanical strength of the optical fiber in the places where the FBG is inscribed. This significantly narrows the possibilities and scope of the FBG arrays. Moreover, the number of FBGs in such an array is limited. This disadvantage is devoid of the method of femtosecond inscribing by laser radiation in the visible, or IR, range. In this case, the removal of the coating is not required, which is noted in [7,8]. However, both methods for manufacturing large FBG arrays require significant, often unjustified data, labor costs.

The solution to the problem was the development of special OFs with extended FBG arrays, which are inscribed directly during the process of extracting the OF with UV radiation [9–15]. FBGs are inscribed in such an optical fiber using pulsed radiation from an excimer UV laser and a phase mask. The number of FBGs per 100 m of such an OF can reach 10,000 pieces. In other words, such OFs can be totally inscribed by FBGs. The increase in the return signal, compared to the Rayleigh scattering level (contrast), reaches values of 50 dB at the Bragg wavelength of 1550 nm. The typical width of the reflection spectrum of the 100%-filling FBG array that is inscribed during the drawing process is 0.3 nm. Using a chirped phase mask to write an FBG array, it is possible to obtain a total reflectance spectrum width of an array of 4 nm. This is necessary for the application of FBG arrays in coherent reflectometry systems operating in wide temperature ranges [15,16]. However, in this case, the level of the reflected signal may decrease.

As compared with the Rayleigh backscattering in optical fibers, backscattering in an OF with an FBG array is characterized by a fairly narrow width of the reflection spectrum, which is determined by the width of the reflection spectrum of individual FBGs. However, within the reflection line itself, the reflection spectra of the OFs with an FBG array are similar to the Rayleigh scattering spectra. They are characterized by a random (Rayleigh) distribution of the amplitude reflection coefficient over the frequency and represent an alternation of maxima and minima on a typical scale, determined by the inverse time of light travel along the entire length of a fiber. Thus, provided that the operating frequency range is within the reflection spectrum of the gratings, OFs with an FBG array are a good alternative to traditional Rayleigh fibers and, therefore, called "artificial Rayleigh fibers". In particular, they can be effectively used in such applications as coherent reflectometry [15,16] and random lasers [17–26].

The spectrum of the reflection of each FBG in array depends on the ambient temperature, tension and refraction coefficient (which is non-permanent in preform and can change a little bit in the OF drawing process). Such factors form a non-phase array with a typical width of 0.2–0.3 nm in opposition to the narrow-width phase grating using in DFB lasers.

Weakly reflective FBG arrays can be also inscribed in active OFs doped with rare-earth ions (erbium, ytterbium and bismuth). It makes it possible to create OFs that combine both an increased reflectivity and the possibility of forming dynamic gratings, which are important for the spectral selection of laser radiation [19,20,27,28]. Such OFs are a good fit for making the cavities for compact random lasers.

The main advantage of random lasers with a cavity, based on an array of FBGs, is the narrow band emission with continuous-wave operation at relatively small cavity lengths (less than 10 m). However, previously such lasers' configurations have shown a relatively low efficiency of the laser's radiation (less than 10%) [20,21,26].

This article is the evolution of our previous works, including an optimization of random laser cavities, based on artificial Rayleigh fibers, and adapted to the conventional telecom wavelength range (also known as "C-band" range). The main novelty is the development of an artificial Rayleigh fiber on the basis of a special photosensitive Er-doped OFs with a germanophosphosilicate core matrix.

2. Materials and Methods

The experimental setup developed "in-house" [26] was implemented to fabricate FBG arrays during the fiber drawing process. The source of UV radiation was an Optosystems CL-5100 pulsed KrF excimer laser [29] with the 248 nm illumination wavelength and a pulse duration of 10 ns. Each of the FBGs in the array was inscribed with a pulse fluence of 400 mJ/cm^2.

In this work, we used a phase mask, manufactured by Ibsen Photonics with a working area of 10 $\times$ 10 mm and a period of 1070 nm. The OF drawing rate is about 10 m/min. The single FBG inscription in the array was performed in one pulse. The phase mask was installed in the immediate vicinity of the fiber being drawn. The typical length of a single FBG was 10 mm long, which has been determined by the width of the laser beam and the working length of the phase mask. The duty cycle of the FBG along the length of the fiber (inscribing density) was controlled by synchronizing the drawing speed and the repetition rate of the UV laser pulses. Thus, with a pulse repetition rate of 10 Hz and an extraction speed of 6 m/min, a fully filled (100%) array with separated FBGs has been achieved. It should be noted that a halving of the laser's pulse repetition frequency (to 5 Hz) at the constant drawing speed indicates the 50% filling of FBG array. In this case, the distance between the individual FBGs is equal to their length, i.e., equal to 10 mm.

In this work, the artificial Rayleigh fiber samples were fabricated from the preform with an erbium-doped core. This preform was synthesized, utilizing the original MCVD method of the separate component deposition, which has been already described in detail in our previous work [26]. The core material was germanophosphosilicate glass doped by the erbium oxide as the activator (Er_2O_3-GeO_2-P_2O_5-SiO_2). According to the results of studying the elemental composition of the core glass using an X-ray analyzer (JEOL 5910LV), the average value of the P_2O_5 concentration was 12 mol.% and GeO_2 was 1.5 mol.%. The erbium concentration turned out to be below the detection limit of the analyzer (less than 0.1 at .%). The preform was stretched and jacketed to the first mode cutoff wavelength of 0.94 μm. The difference between the refractive indices of the core and cladding in the jacketed preform, measured using a Photon Kinetics P2610 preform analyzer, was 0.015.

We used the method of optical frequency domain reflectometry (OFDR) [30] with a Luna 4400 device to characterize the fabricated FBG arrays. In addition, we used a Yokogawa AQ6370D optical spectrum analyzer (OSA) with a maximum optical resolution of 0.02 nm for the spectral measurements of the FBG array reflection. Additionally, this OSA was used for the measurements of the OF absorption spectra. The source of radiation for the spectral measurements was a superluminescent diode with a fiber output (for the FBG array samples) or a halogen lamp (for the conventional OF samples).

The single-mode OF with a standard outer diameter of 125 μm was drawn from the preform. The fiber had a core diameter of 4.8 μm and cutoff wavelength of 900 nm.

The level of background (or "gray") optical losses was about 2 dB/km, which is a sign of the homogeneous structure of the core glass and the absence of the transition metal impurities in it. The intensities of the Er^{3+} ion absorption peaks at 980, 1490 and 1535 nm in the fiber were 3.3, 3.1 and 7 dB/m, respectively. The transmission spectrum of the fiber in the extended telecommunications wavelength range (1460–1620 nm) measured with an optical resolution of 1 nm is shown in Figure 1.

The small-signal gain of an active fiber is fundamentally important to find the optimal configuration of a fiber laser cavity and evaluate its generation characteristics. The small-signal gain in the fiber was measured in the wavelength range of 1525–1610 nm using two APEX AP1000-5 and AP1000-8 platforms equipped with AP3350A tunable semiconductor single-frequency lasers (operated in range of 1525–1610 nm) and AP3352A (operated in range of 1575–1610 nm). Both were used as weak signal sources. Two laser diodes radiated at a wavelength of 976 or 1485 nm were used as sources for the active fiber pumping. Using the technique described above, the spectral dependences of the gain in the telecommunications wavelength range (C- and L-range) were obtained for two pumping schemes: at a wavelength of 976 nm (Figure 2, curve "1") and 1485 nm (Figure 2, curve "2").

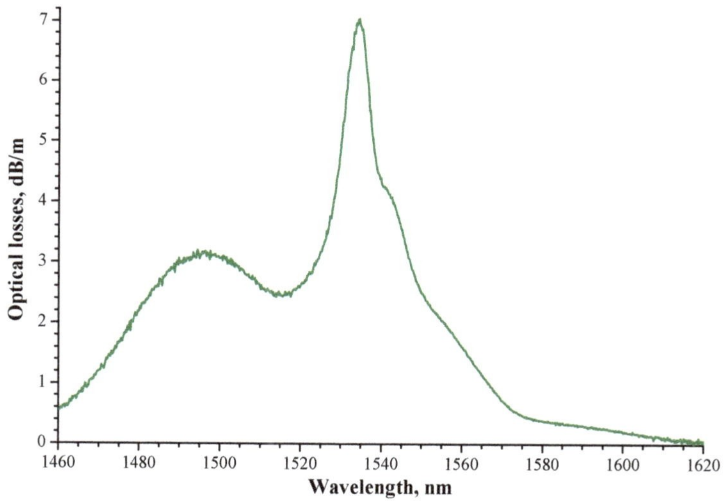

Figure 1. Transmission spectrum of the optical fiber in the wavelength range of 1460–1620 nm.

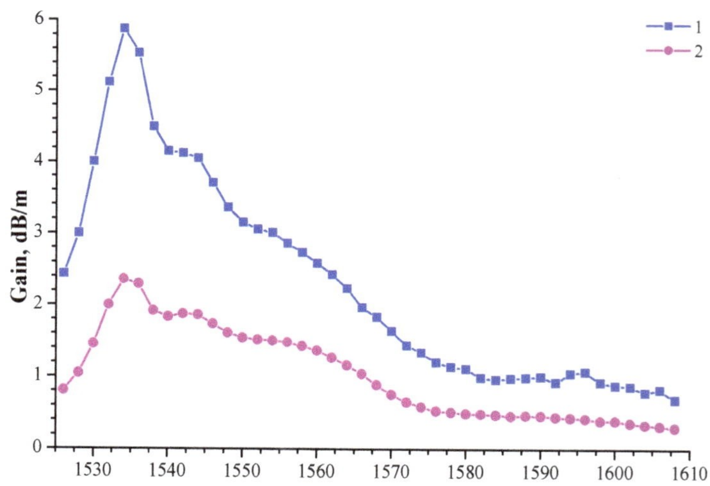

Figure 2. Spectral dependences of the small-signal gain in the wavelength range 1525–1610 nm, measured in the fiber pumping mode at a wavelength of 976 nm ("1") and 1490 nm ("2").

The main object of our research in this work was the artificial Rayleigh fiber: a 7-m-long array filled on the 100% uniformly by weakly reflective FBGs. Moreover, another similar fiber sample with a 50% FBG filling was fabricated for inscription contrast measurement (the excess of the return signal above the Rayleigh level). The FBG inscription contrast obtained by the OFDR technique was up to 45 dB (see Figure 3).

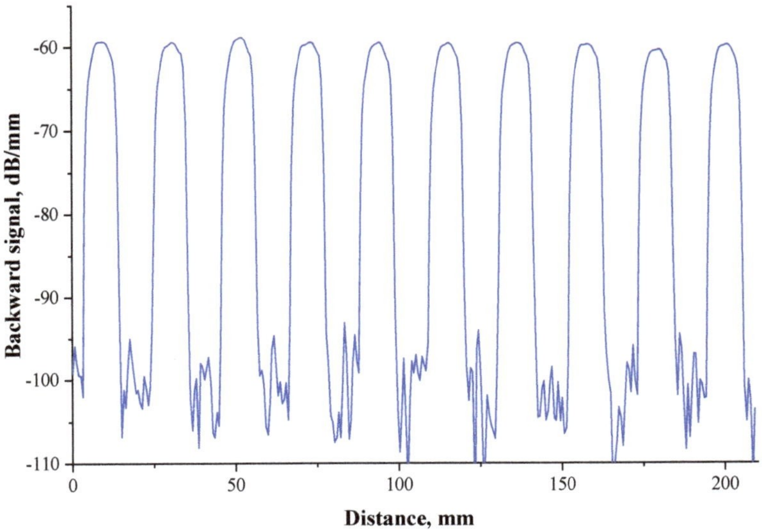

Figure 3. Frequency reflectogram (OFDR) of the experimental 50%-filled weakly reflective FBG array with a central reflection wavelength at 1548 nm.

Figure 4 reveals the optical reflection spectrum of the 100%-filled 7-m-long weakly reflective FBG array, which has been obtained by OSA with an optical resolution of 0.1 nm. Next, this sample was used as the "random" laser cavity.

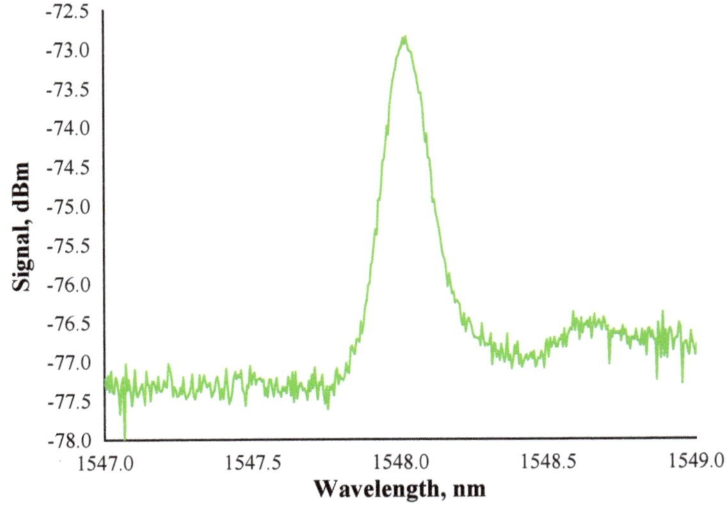

Figure 4. Optical reflection spectrum of the 7-m-long weakly reflective FBG array.

The basic scheme of the laser investigation experimental setup is depicted in Figure 5. The "random" laser cavity (1) was backward-pumped via a wavelength-division multiplexer (WDM) (2) by using one of the two laser diodes (3): Photonics 3S1999CHP (emission wavelength 974.5 nm) and Anritsu GF4B701 (emission wavelength 1485 nm). The optical power of the pumping radiation was controlled by a CNILaser LP100 powermeter. The laser output radiation passed through the optical isolator (4) to avoid the influence of back-reflection. The OSA Yokogawa AQ6370D (5) was used for the laser emission spectral measurements, and Grandway FHM2B01 or JDSU MP-60 powermeters (6) were used for the laser optical power measurements. Optionally, the highly reflective FBG (7) was used to extend the "random" laser cavity and its lasing properties have also been studied. In addition to the basic measurements, the laser operation mode has been investigated by Extech MS6060 oscilloscope and the special self-heterodyne radio-frequency (RF) setup.

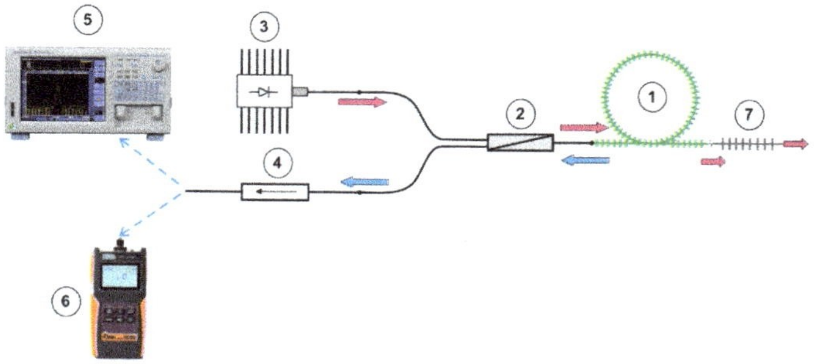

Figure 5. The scheme of the experimental setup: 1– laser cavity; 2–wavelength-division multiplexer (WDM); 3–laser diodes; 4–optical isolator; 5–optical spectrum analyzer; 6–optical powermeters; 7–highly reflective FBG.

3. Results

The cavity of the "random" laser studied in this work generated narrow-band radiation at a wavelength of 1548 nm. As can be seen from Figure 6, the "telecom wavelength-range" spectrum of the pumped 7-m-long artificial Rayleigh fiber has a single narrow peak, while only the erbium amplified luminescence (ASE) is visible in the spectrum of the same conventional (i.e., without weakly reflective FBGs) pumped fiber.

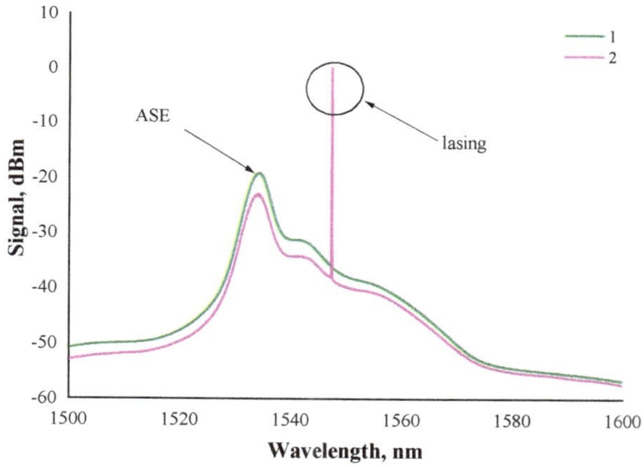

Figure 6. Emission spectra of the conventional (i.e., non-containing weakly reflective FBG array) active OF ("1") and the artificial Rayleigh fiber ("2"), obtained at the same power level (~30 mW) of the 976 nm wavelength pump. "ASE"—spontaneous emission, "lasing"—laser generation.

One of the most important characteristics of a fiber laser is the dependence of the output radiation power from the power of the pump radiation coupled into the cavity. Figure 7 shows the power dependencies for the 974.5 nm-wavelength pumping case. The graph "1" in Figure 7 attributes to the laser output power depending on the pump radiation coupled power. The second graph "2" reveals the unabsorbed pump power (i.e., pump radiation passed through the cavity) versus the coupled pump power.

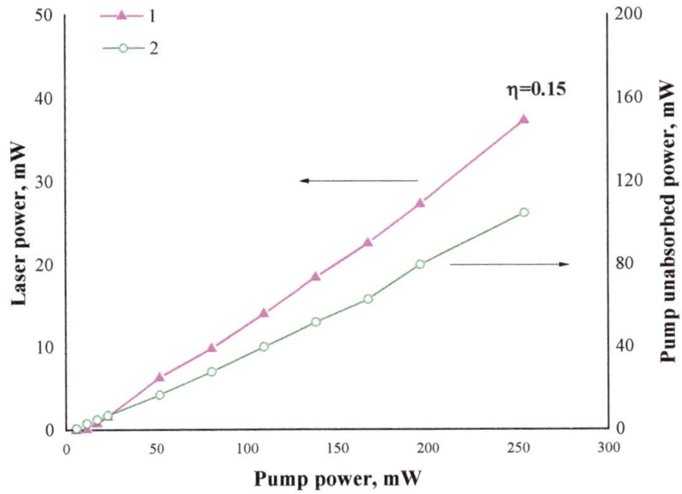

Figure 7. Output power ("1") of the "random" laser cavity and the unabsorbed pump power ("2") versus the coupled pump power with a wavelength of 974.5 nm. Pink line–left scale. Black line right scale.

The analysis of Figure 7 gives us a result of the differential power efficiency (or slope efficiency, η) of the "random" laser: 0.15. It should be noted that a rather significant part (~40%) of the pump radiation was not absorbed in the cavity and therefore the η value was not maximum.

A relatively simple and convenient way for increasing the slope efficiency of a "random" cavity without the risk of lasing instability escalation is the extension of this cavity by the single separate highly reflective FBG [25]. The reflection peak of such FBGs in this case must be wavelength matched exactly with the reflection maximum of the weakly reflective FBG array (i.e., lasing wavelength), which form a basic "random" cavity. The FBG used in the present work had a length of 10 mm and was inscribed in a photosensitive (germanosilicate) OF, with waveguide characteristics close to the parameters of the studied active OF: the cutoff wavelength was 910 nm, the refractive index difference was 0.014. This FBG with a 90% reflection on the wavelength of 1548 nm was spliced to the 7-m-long "random" cavity from the side opposite to the direction of launch the pump radiation. The optical loss at the splicing point did not exceed 0.1 dB.

As seen in the Figure 8 (curves "1" and "2"), the modification "random" cavity by the 90%-reflective FBG did not lead to any significant changes in the shape of the emission spectrum. The slight difference in the peak wavelength of the "1" and "2" spectra may be due to the temperature effects.

On the other hand, the "random" cavity modified with a 90%-reflective FBG demonstrated a twofold increase in the output power and slope efficiency (η = 0.33) at the same values of the pump power (see Figure 9, graph "1"). This can be explained by the more efficient absorption of the pump radiation in the cavity modified by the FBG (compare graphs "2" in the Figures 7 and 9), despite the fact that its length (7 m) did not change much. It is extremely important to point out that lasing with a 90%-reflective FBG was unreachable in configuration with a conventional 7-m-long active OF. The lasing was possible only when this FBG was attached to the array of weakly reflective FBGs, i.e., segment of the artificial Rayleigh fiber.

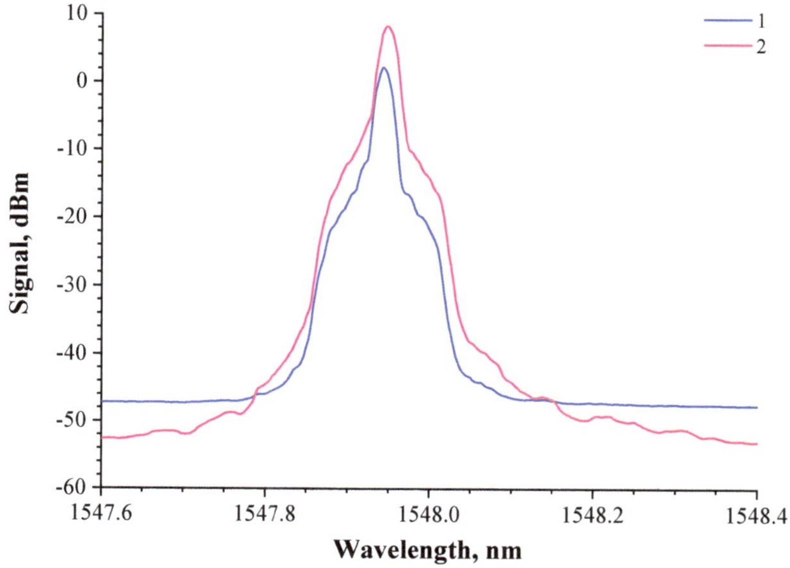

Figure 8. Emission spectra of the "random" laser cavities obtained at a 974.5 nm wavelength pumping (coupled power of 30 mW): the pristine cavity (1) and the same cavity modified by the 90%-reflective FBG (2).

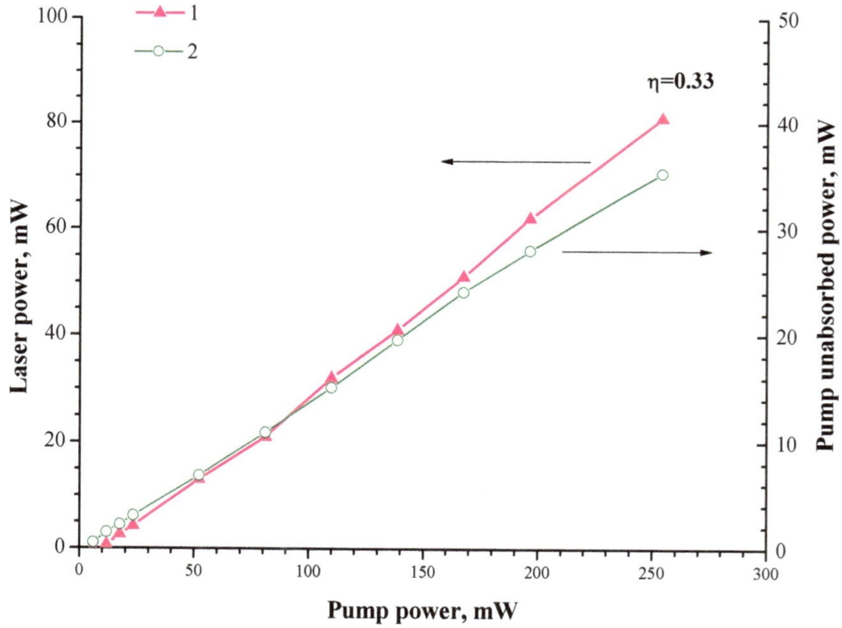

Figure 9. Output power ("1") of the "random" laser cavity, modified with FBG, and the unabsorbed pump power ("2") versus the coupled pump power with a wavelength of 974.5 nm. Red line—left scale. Green line—right scale.

In the case of 1485 nm wavelength pumping, the influence of the cavity extension by the 90%-reflective FBG was much more significant. Importantly, the lasing threshold in the "pristine" 7-m-long cavity (without the 90%-reflective FBG) was not achieved at any pumping power level. On the contrary, as shown in the Figure 10, the same laser cavity modified by the 90%-reflective FBG demonstrated a good lasing threshold (~10 mW of pumping power) and slope efficiency (η = 0.29), which was comparable in value to the slope efficiency at the 974.5 nm wavelength pumping (η = 0.33).

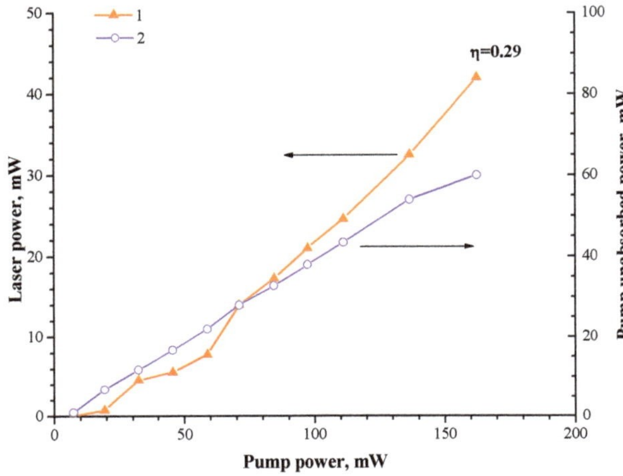

Figure 10. Output power ("1") of the "random" laser cavity, modified with FBG, and the unabsorbed pump power ("2") versus the coupled pump power with a wavelength of 1485 nm. Orange line—left scale. Violet line—right scale.

The long-term laser power stability has been studied. The laser cavity extended by the 90%-reflective FBG was pumped at the 974.5 nm wavelength with a 50 mW input power level. Figure 11 displays the laser's output power value with a long time duration. It can be concluded that the laser power is sufficiently stable at a constant external temperature (295 K).

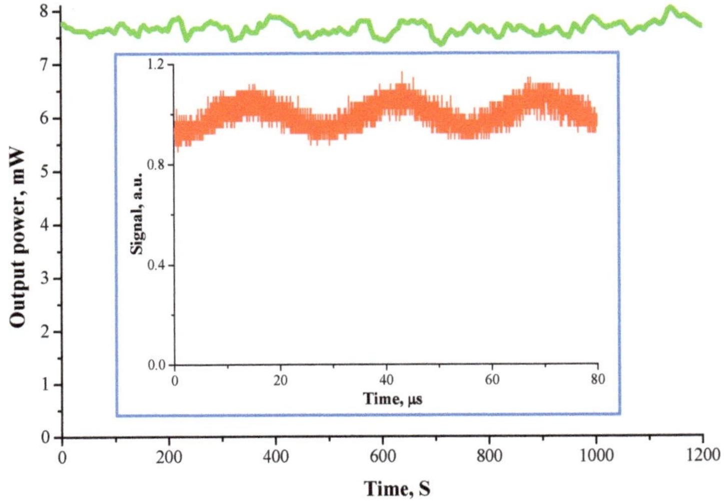

Figure 11. Time dependence of the laser's output power (green). Inset—oscilloscope trace (red).

It was also found that the studied laser operated strictly in a continuous-wave (CW) mode, without any self-pulsing (such as a Q-switch). This result is in good agreement with our previous works [19–22,25]. The CW laser operation can be explained by the contribution of the dynamic grating formation [19,20]. On the other hand, one can see in the inset of Figure 11 a weak oscillation with a frequency of 35–40 kHz. This behavior and the oscillation frequency value, however, are typical for Er-doped fiber lasers and it was investigated earlier in publications [21,31,32].

Figure 12 shows the resulting spectrum of the laser emission linewidth measurement by using the self-heterodyne method [20,33]. This measurement was carried out via a radio-frequency (RF) FSH8 Rohde and Schwarz spectrum analyzer connected with an unbalanced Mach–Zehnder interferometer involving the 40 MHz frequency electro-optic modulator and a 50-km-long fiber delay line. The laser was pumped at the 976 nm wavelength with a power level of 72 mW. The shown RF spectrum has resulted of 20 times averagement of the laser output signal during the scanning time of 374 ms. The measured RF spectrum has a good fit with a Lorentz function and so the linewidth value was determined as about of 550 Hz at the −3 dB intensity level.

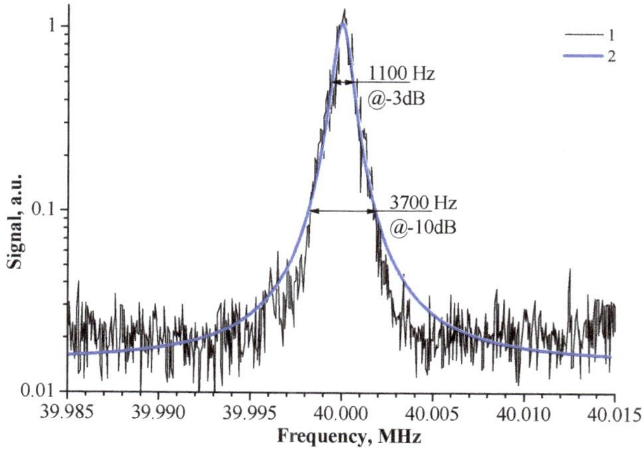

Figure 12. Averaged RF spectrum ("1") and fitted by the Lorentz function ("2").

4. Discussion

For the first time, comparative studies of the random laser efficiency with a cavity based on the artificial Rayleigh fibers in pumping modes at wavelengths of 974.5 and 1485 nm have been carried out. Initially, for the experiments, the OF cavity length was chosen as 7 m by a balance between the FBG array reflection and erbium ion absorption. When using a source with a wavelength of 974.5 nm in the cavity with a length of 7 m, it leads to the appearance of lasing at a wavelength of 1548 nm at a rather low threshold value of an injected pump power, about 7 mW. On the contrary, in the same cavity, when it was pumped by the source with a wavelength of 1485 nm, the lasing threshold would not be reached. In the cavity modified with a 90% FBG, the laser generation was obtained in the pumping mode at the wavelength of 1485 nm. The slope efficiency of the modified "random" cavity in the pumping modes at wavelengths of 974.5 and 1485 nm turned out to be comparable: 0.33 and 0.29, respectively.

The effect of the significant (many times) increase in a laser efficiency in a "random" cavity, modified by adding a single FBG with a reflection coefficient of 90%, has been found. Such an effect can be explained by changing the lasing configuration from open to half-open cavity configuration [34] that increases a length of laser's cavity, returning lasing power from "forward" direction to "backward" direction and finally reducing lasing threshold. The absence of lasing at 1485 nm without 90% FBG can be explained by non-optimal pumping conditions, because the fiber's cutoff wavelength is 900 nm, thus, it means that the pumping process of the active core at 1485 nm is a multimode. Once again, it clearly demonstrates the advancement of the highly reflective FBG extension in such laser configurations. This FBG also functions as a highly efficient wavelength selector in the case of heterogeneity in the length of the artificial Rayleigh fiber, as noted earlier [25].

The effect of long-time changes in the level of the output signal shown in Figure 11 is explained by not using the polarization maintaining items to produce such a laser. It causes problems of long-time lasing stability. Our next step will be a random laser based on the polarization maintaining items (include a single polarization artificial Rayleigh fiber) to prevent such problems in the future. A later optimization conversion efficiency is possible by the optimal concentration of the erbium ions, cavity length, FBG contrast (a reflection level of artificial Rayleigh fiber) and the wavelength of lasing (for example, 1530 nm is the wavelength of the maximum amplification).

The investigated laser scheme operates in continuous-wave mode. Such a continuous lasing mode is typical for random lasers [2]. The measured lasing single-frequency emission linewidth is about 550 Hz, which fits with our previous works [20,21]. Such a result is opposite to earlier works concerning many FBGs inscribed in the erbium doped OF [5], where multiple modes of lasing take place. The difference explained by that cavity in our case consists of many FBGs that totally coat OF and form a continuous reflection track similar to the Rayleigh scattering, instead of a set of discrete FBGs (point reflectors) as were previously used in pioneering works. Single frequency lasing is also explained by the contribution of dynamic grating formation, which possibly support the main lasing mode and operates as a dynamic filter—as described in our previous publications [19,20].

5. Conclusions

It has been demonstrated that an erbium-doped fiber laser with a "random" resonator is capable of operating at room temperature in a continuous-wave mode for a long time (at least tens of minutes), which is extremely important from the point of view of the prospects for its use as a fiber source of high coherence optical radiation. The feature of the described laser configurations is a high efficiency (up to 33%) with a combination of narrow linewidth (~550 Hz) radiation.

The proposed random laser is a simple, compact, cost-effective and highly efficient solution for many practical applications such as telecommunication and fiber-optic sensing.

Author Contributions: Conceptualization, S.P., A.R. and D.L.; methodology, S.P., V.V., A.K., I.V. and V.I.; software, D.R. and S.P.; validation, Y.C.; investigation, A.R., S.P., A.B., A.S. and D.R.; data curation, S.P. and Y.C.; writing—original draft preparation, A.R. and S.P.; supervision, A.R. and S.P.; project administration, O.B.; funding acquisition, O.B.; Writing—review and editing, S.P. All authors have read and agreed to the published version of the manuscript.

Funding: The work was carried out within the framework of the Kotelnikov IRE RAS state task.

Institutional Review Board Statement: Not applicable.

Informed Consent Statement: Not applicable.

Data Availability Statement: Not applicable.

Conflicts of Interest: The authors declare no conflict of interest.

References

1. Fotiadi, A.A.; Kiyan, R.V. Cooperative stimulated Brillouin and Rayleigh backscattering process in optical fiber. *Opt. Lett.* **1998**, *23*, 1805–1807. [CrossRef]
2. Turitsyn, S.; Babin, S.; El-Taher, A.; Harper, P.; Churkin, D.; Kablukov, S.; Ania-Castañón, J.; Karalekas, V.; Podivilov, E. Random distributed feedback fibre laser. *Nat. Photonics* **2010**, *4*, 231. [CrossRef]
3. Fotiadi, A.A. Random lasers: An incoherent fibre laser. *Nat. Photonics* **2010**, *4*, 204–205. [CrossRef]
4. Turitsyn, S.K.; Babin, S.A.; Churkin, D.V.; Vatnik, I.D.; Nikulin, M.; Podivilov, E.V. Random distributed feedback Fibre lasers. *Phys. Rep.* **2014**, *542*, 133–193. [CrossRef]
5. Lizárraga, N.; Puente, N.P.; Chaikina, E.I.; Leskova, T.A.; Méndez, E.R. Single-mode Er-doped fiber random laser with distributed Bragg grating feedback. *Opt. Express* **2009**, *17*, 395–404. [CrossRef] [PubMed]
6. Skvortsov, M.I.; Abdullina, S.R.; Vlasov, A.A.; Zlobina, E.A.; Lobach, I.A.; Terentyev, V.S.; Babin, S.A. FBG array-based random distributed feedback Raman fibre laser. *Quantum Electron.* **2017**, *47*, 696. [CrossRef]

7. Bronnikov, K.; Wolf, A.; Yakushin, S.; Dostovalov, A.; Egorova, O.; Zhuravlev, S.; Semjonov, S.; Wabnitz, S.; Babin, S. Durable shape sensor based on FBG array inscribed in polyimide-coated multicore optical fiber. *Opt. Express* **2019**, *27*, 38421. [CrossRef]
8. Przhiialkovskii, D.; Butov, O. High-precision point-by-point fiber Bragg grating inscription. *Results Phys.* **2021**, *30*, 104902. [CrossRef]
9. Askins, C.G.; Putnam, M.A.; Williams, G.M.; Fiebele, E.J. Stepped-wavelength optical-fiber Bragg grating arrays fabricated in line on a draw tower. *Opt. Lett.* **1994**, *19*, 147–149. [CrossRef]
10. Guo, H.; Tang, J.; Li, X.; Zheng, Y.; Yu, H.; Yu, H. On-line writing identical and weak fiber Bragg grating arrays. *Chin. Opt. Lett.* **2013**, *11*, 030602.
11. Zaitsev, I.; Butov, O.; Voloshin, V.; Vorob'ev, I.; Vyatkin, M.; Kolosovskii, A.; Popov, S.; Chamorovskii, Y. Optical Fiber with Distributed BraggType Reflector. *J. Comm. Technol. Elec.* **2016**, *61*, 639. [CrossRef]
12. Popov, S.; Butov, O.; Kolosovski, A.; Voloshin, V.; Vorob'ev, I.; Vyatkin, M.; Fotiadi, A.; Chamorovski, Y. Optical Fibres with Arrays of FBG: Properties and Application. In Proceedings of the PIERS, IEEE Xplore, 1568, St. Petersburg, Russia, 22–25 May 2017.
13. Chamorovskiy, Y.; Butov, O.; Kolosovskiy, A.; Popov, S.; Voloshin, V.; Vorob'ev, I.; Vyatkin, M. Metal-coated Bragg grating reflecting fibre. *Opt. Fiber Technol.* **2017**, *34*, 30. [CrossRef]
14. Chamorovskiy, Y.; Butov, O.; Kolosovskiy, A.; Popov, S.; Voloshin, V.; Vorob'ev, I.; Vyatkin, M.; Odnobludov, M. Long tapered fiber with array of FBG. *Opt. Fiber Technol.* **2019**, *50*, 95. [CrossRef]
15. Popov, S.; Butov, O.; Kolosovskii, A.; Voloshin, V.; Vorob'ev, I.; Isaev, V.; Vyatkin, M.; Fotiadi, A.; Chamorovsky, Y. Optical fibres and fibre tapers with an array of Bragg gratings. *Quantum Electron.* **2019**, *49*, 1127. [CrossRef]
16. Kharasov, D.; Bengalskii, D.; Vyatkin, M.; Nanii, O.; Fomiryakov, E.; Nikitin, S.; Popov, S.; Chamorovsky, Y.; Treshchikov, V. Extending the operation range of a phase-sensitive optical time-domain reflectometer by using fibre with chirped Bragg gratings. *Quantum Electron.* **2020**, *50*, 510. [CrossRef]
17. Popov, S.; Chamorovsky, Y.; Mégret, P.; Zolotovskii, I.; Fotiadi, A. Brillouin Random Lasing in Artifice Rayleigh Fiber. In Proceedings of the ECOC, IEEE Xplore, 1, Valensia, Spain, 27 September–1 October 2015.
18. Popov, S.; Butov, O.; Chamorovskiy, Y.; Isaev, V.; Kolosovskiy, A.; Voloshin, V.; Vorob'ev, I.; Vyatkin, M.; Mégret, P.; Odnoblyudov, M.; et al. Brillouin lasing in single-mode tapered optical fiber with inscribed Fiber Bragg Grating Array. *Results Phys.* **2018**, *9*, 625. [CrossRef]
19. Popov, S.; Butov, O.; Chamorovski, Y.; Isaev, V.; Mégret, P.; Korobko, D.; Zolotovskii, I.; Fotiadi, A. Narrow linewidth short cavity Brillouin random laser based on Bragg grating array fiber and dynamical population inversion gratings. *Results Phys.* **2018**, *9*, 806–808. [CrossRef]
20. Popov, S.; Butov, O.; Bazakutsa, A.; Vyatkin, M.; Chamorovskii, Y.; Fotiadi, A. Random lasing in a short Er-doped artificial Rayleigh fiber. *Results Phys.* **2020**, *16*, 102868. [CrossRef]
21. Popov, S.M.; Butov, O.V.; Bazakutsa, A.P.; Vyatkin, M.Y.; Chamorovskiy, Y.K.; Panajotov, K.; Fotiadi, A.A. Narrow linewidth random laser based on short Er-doped artifice Rayleigh fiber. *Proc. SPIE* **2020**, *11357*, 318–327.
22. Rybaltovsky, A.; Popov, S.; Lipatov, D.; Umnikov, A.; Abramov, A.; Morozov, O.; Ryakhovskiy, D.; Voloshin, V.; Kolosovskii, A.; Vorob'ev, I.; et al. Photosensitive Yb-doped germanophosphosilicate artificial Rayleigh fibers as a base of random lasers. *Fibers* **2021**, *9*, 53. [CrossRef]
23. Skvortsov, M.I.; Wolf, A.A.; Dostovalov, A.V.; Egorova, O.N.; Semjonov, S.L.; Babin, S.A. Narrow-Linewidth Er-Doped Fiber Lasers With Random Distributed Feedback Provided By Artificial Rayleigh Scattering. *J. Light. Technol.* **2022**, *6*, 1829–1835. [CrossRef]
24. Skvortsov, M.I.; Abdullina, S.R.; Wolf, A.A.; Dostovalov, A.V.; Churin, A.E.; Egorova, O.N.; Semjonov, S.L.; Proskurina, K.V.; Babin, S.A. Single-frequency erbium-doped fibre laser with random distributed feedback based on disordered structures produced by femtosecond laser radiation. *Quantum Electron.* **2021**, *51*, 1051–1055. [CrossRef]
25. Rybaltovsky, A.; Popov, S.; Ryakhovskiy, D.; Abramov, A.; Umnikov, A.; Medvedkov, O.; Voloshin, V.; Kolosovskii, A.; Vorob'ev, I.; Chamorovskiy, Y.; et al. Random Laser Based on Ytterbium-Doped Fiber with a Bragg Grating Array as the Source of Continuous-Wave 976 nm Wavelength Radiation. *Photonics* **2022**, *9*, 840. [CrossRef]
26. Popov, S.M.; Butov, O.V.; Kolosovskii, A.O.; Voloshin, V.V.; Vorob'ev, I.L.; Isaev, V.A.; Ryakhovskii, D.V.; Vyatkin, M.Y.; Rybaltovskii, A.A.; Fotiadi, A.A.; et al. Optical fibres with an inscribed fibre Bragg grating array for sensor systems and random lasers. *Quantum Electron.* **2021**, *51*, 1101–1106. [CrossRef]
27. Spirin, V.; López-Mercado, C.; Kinet, D.; Mégret, P.; Zolotovskiy, I.; Fotiadi, A. Single longitudinal-mode brillouin fiberlaser passively stabilized at pump resonance frequency with dynamic population inversion grating. *Laser Phys. Lett.* **2013**, *10*, 015102. [CrossRef]
28. Lobach, I.A.; Drobyshev, R.V.; Fotiadi, A.A.; Podivilov, E.V.; Kablukov, S.I.; Babin, S.A. Open-cavity fiber laser with distributed feedback based on externally or self-induced dynamic gratings. *Opt. Lett.* **2017**, *42*, 4207–4210. [CrossRef]
29. Web-Site of Optosystems. Available online: https://optosystems.ru/en/product/cl-5000/ (accessed on 22 May 2023).
30. Soller, B.; Gifford, D.; Wolfe, M.; Froggatt, M. High resolution optical frequency domain reflectometry for characterization of components and assemblies. *Opt. Express* **2005**, *13*, 666. [CrossRef]
31. Smirnov, A.M.; Bazakutsa, A.P.; Chamorovskiy, Y.K.; Nechepurenko, I.A.; Dorofeenko, A.V.; Butov, O.V. Thermal Switching of Lasing Regimes in Heavily Doped Er3+ Fiber Lasers. *ACS Photonics* **2018**, *5*, 5038–5046. [CrossRef]

32. Skvortsov, M.; Wolf, A.; Dostovalov, A.; Vlasov, A.; Akulov, V.; Babin, S. Distributed feedback fiber laser based on a fiber Bragg grating inscribed using the femtosecond point-by-point technique. *Laser Phys. Lett.* **2018**, *15*, 035103. [CrossRef]
33. Richter, L.E.; Mandelberg, H.I.; Kruger, M.S.; Mcgrath, P.A. Linewidth determination from self-heterodyne measurements with subcoherence delay times. *IEEE J. Quantum Electron.* **1986**, *22*, 2070–2074. [CrossRef]
34. Zhang, W.L.; Rao, Y.J.; Zhu, J.M.; Wang, Z.X.; Jia, X.H. Low threshold 2nd-order random lasing of a fiber laser with a half-opened cavity. *Opt. Express* **2012**, *20*, 14400–14405. [CrossRef] [PubMed]

photonics

MDPI

Communication

Relocking and Locking Range Extension of Partially Locked AMLL Cavity Modes with Two Detuned RF Sinusoids

Shree Krishnamoorthy [1,2,*,†] and Anil Prabhakar [1]

1 Indian Institute of Technology Madras, Chennai 600036, India; anilpr@iitm.ac.in
2 National Center for Biological Sciences, Bengaluru 560065, India
* Correspondence: shree.krishnamoorthy@tyndall.ie
† Current address: Tyndall National Institute, T12 R5CP Cork, Ireland.

Abstract: Actively mode-locked fiber ring lasers (AMLLs) with loss modulators are used to generate approximately 100 ps pulses with 100 MHz repetition. RF detuning around the fundamental frequency, f_0, causes a loss in phase lock (unlocking) of cavity modes and partial mode locking. Multiple RF inputs are shown, theoretically, to relock and extend the locking range of cavity modes in a detuned partially mode-locked AMLL. A custom-built Yb^{3+}-doped AMLL with $f_0 = 26$ MHz, and operating wavelength of 1064 nm, was used to experimentally verify the theoretical predictions. Two RF sinusoidal signals with constant phase and equal amplitude resulted in an extension of the range by $X_n = 6.4$ kHz in addition to the range $R_n = 14.34$ kHz with single input for the mode $n = 10$. An increase in locking range was also observed for higher modes. Pulsewidth reduction to approximately 205 ps from about 2 ns was also observed in the AMLL.

Keywords: fiber lasers; mode locking; active mode locking; partial mode locking; laser cavity resonators; laser mode locking; laser tuning; ring lasers; laser theory; radio frequency; detuned mode locking; mode relocking

1. Introduction

Actively mode-locked fiber ring lasers (AMLLs) produce ultrashort pulses with high peak powers and can be synchronized with other light sources for both advanced microscopy and spectroscopic techniques [1–5]. AMLLs have been used to generate both radio frequency (RF) and optical pulsetrains from a laser cavity [6–8]. The versatility of AMLLs for both electrical and ultrashort optical pulses with low maintenance and few stability issues makes them a good candidate for adoption into large systems as synchronized, pulsed multiple laser sources.

Traditionally, pulse formation in an AMLL occurs through the process of mode locking, where all cavity modes, around the central carrier wavelength, are phase-locked with the RF driving signal, typically a single sinusoid. Upon detuning of the driving signal, either due to drift in RF signal or environmental fluctuations, partial locking of the AMLL occurs. This leads to a broadening of pulses due to lock loss of higher modes in both fundamentally and harmonically mode-locked lasers [9–15]. When subject to a large detuning, the laser operates in either Q-switched or amplitude-modulated regimes with pulsewidths in the order of a few ns. This regime also has a wide range of applications, including pulse generation in Raman lasers [12,16–19]. The current methods used to counter detuning are by either modifying the AMLL cavity or using additional optical components. Both of these approaches lead to undesirable changes in the quality of pulses from the laser in terms of synchronization with the source, the pulse's optical wavelength and repetition rate [20–24]. For a successful adoption of AMLLs as optical pulsed sources in synchronized systems, we must develop novel methods to counter the effects of detuning.

Multiple optical injection signals have been used in semiconductor lasers, where the system dynamics have been studied through numerical techniques [25–28]. In such systems,

Citation: Krishnamoorthy, S.; Prabhakar, A. Relocking and Locking Range Extension of Partially Locked AMLL Cavity Modes with Two Detuned RF Sinusoids. *Photonics* **2023**, *10*, 735. https://doi.org/10.3390/photonics10070735

Received: 27 April 2023
Revised: 11 June 2023
Accepted: 14 June 2023
Published: 27 June 2023

there is an improvement in the time bandwidth product, optical linewidth and timing jitter in the laser [27]. Further, multiple injection signals have shown the occurrence of typical Adler-type and atypical Arnold-type locking regimes [28,29]. In this article, we extend our previous work [15,30,31], both theoretically and experimentally, to include multiple RF signals. The use of two RF input signals is a novel method of driving a detuned fiber AMLL. We thus demonstrate an improvement in the number of locked modes by the process of *relocking* using this method for the first time in the literature. Further, we provide a novel demonstration of improvement in the pulsewidths of the partially mode-locked laser with two RF input signals without any additional optical components. We demonstrate a new regime of operation for AMLLs, where, temporally, pulse bunches are produced periodically with varying numbers of locked modes in each of the pulses.

2. Theory

Consider the AMLL with multiple loss-modulating RF inputs as shown in Figure 1a. The experimental implementation of the AMLL with two inputs is shown in Figure 1b. The pulse, $P(f)$, interacts with the modulation signal, $M(f)$, to give $I(f)$ at the modulator. The optical construction includes a pump (980 nm) coupled to the gain fiber (Yb) using a wavelength division multiplexer (WDM), a circulator and a fiber Bragg grating at 1064 nm, and a 50/50 optical coupler leading to the optical output. The two inputs considered for the experiments are shown in Figure 1b. In this model, we consider the optical energy in the cavity circulating with a fundamental round-trip frequency of f_r. Periodic loss modulation causes injection of energy into the nth cavity mode at a frequency nf_r, in phase lock with the input signal [30,32–34]. In an ideal mode-locked cavity with fundamental resonant frequency, f_0, the cavity modes occur at frequencies nf_0 and we have $f_r = f_k = f_0$, where f_k is the detuned input frequency.

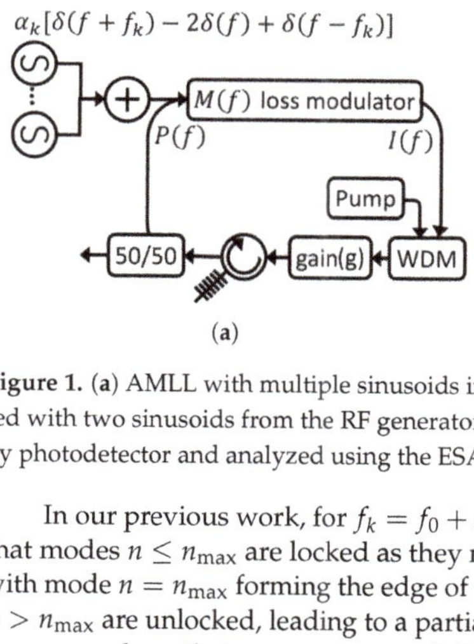

(**a**)

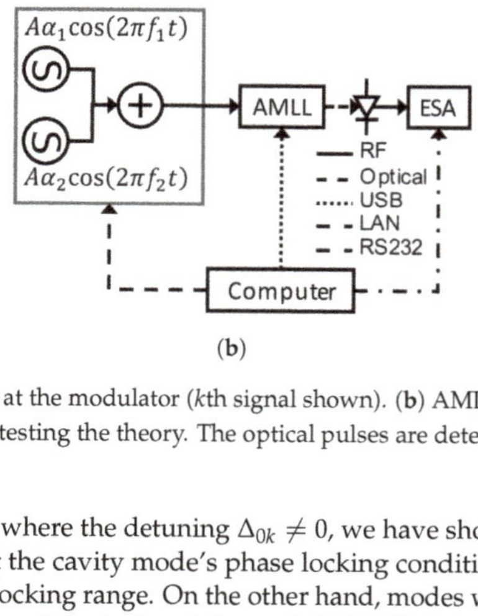

(**b**)

Figure 1. (**a**) AMLL with multiple sinusoids input at the modulator (kth signal shown). (**b**) AMLL is fed with two sinusoids from the RF generator for testing the theory. The optical pulses are detected by photodetector and analyzed using the ESA.

In our previous work, for $f_k = f_0 + \Delta_{0k}$, where the detuning $\Delta_{0k} \neq 0$, we have shown that modes $n \leq n_{\max}$ are locked as they meet the cavity mode's phase locking conditions, with mode $n = n_{\max}$ forming the edge of the locking range. On the other hand, modes with $n > n_{\max}$ are unlocked, leading to a partially mode-locked and detuned AMLL [30]. In this paper, we show that one can successfully recover the phase lock, i.e., *relock*, an unlocked mode by using multiple sinusoidal inputs at the loss modulator. For this, we extend the single sinusoidal input and single circulating pulse detuned AMLL model to one with multiple sinusoidal inputs and multiple circulating pulses in the cavity.

Consider the total pulse, $P(f)$,

$$P(f) = \sum_r P_r(f) \tag{1}$$

where $r \in \mathbb{N}$, and the circulating pulses $P_r(f)$ have a repetition frequency f_r, given by,

$$P_r(f) = \gamma_r \sum_n A_n \delta(f - nf_r). \tag{2}$$

γ_r is the relative strength such that $\sum_r \gamma_r = 1$, A_n is the amplitude of the mode n and $\delta()$ is the Dirac delta function. The detuning of this pulsetrain is $\Delta_{r0} = f_r - f_0$.

The AMLL has gain g, quality factor Q, central optical frequency ν_s, laser linewidth ν_L and cold cavity resonance f_C. Each $P_r(f)$ individually satisfies the mode locking conditions [35]. Following the treatment from our previous work [30], we describe the perturbed AMLL with multiple inputs and multiple pulsetrains as

$$\sum_r \gamma_r \left\{ j\frac{2Qf_C}{\nu_s f_0} n\Delta_{r0} - g\left[-2\frac{n^2 f_0}{\nu_L^2}\Delta_{r0} \right] \right\} A_n = \widetilde{I}_n, \tag{3}$$

where $\widetilde{I}_n$ is the modulated amplitude at mode n and consists of interactions between the pulses and the modulating signal. In this work, we neglect the changes in amplitude A_n and gain g in the detuned AMLL. Further, it is assumed that the detuning in the RF signal does not deviate from the central optical frequency ν_s. In the next section, we shall look at the perturbed AMLL described in (3) closely.

2.1. Interaction of Pulses in the Modulator

As shown in Figure 1a, the total input signal is

$$\mathcal{M}(f) = -M \sum_k \alpha_k [\delta(f + f_k) - 2\delta(f) + \delta(f - f_k)], \tag{4}$$

where the kth sinusoid has frequency f_k and the normalized relative amplitude α_k, such that $\sum_k \alpha_k = 1$, and the modulation depth is M. The total modulating signal, $I(f)$, is obtained by convolving (4) with (1):

$$I(f) = -M \sum_k \sum_r \alpha_k [P_r(f + f_k) - 2P_r(f) + P_r(f - f_k)]. \tag{5}$$

Note the injection between adjacent modes due to the action of the sinusoidal modulation. Any pairs of f_r and f_k are detuned from each other such that

$$f_k = f_r + \Delta_{kr}. \tag{6}$$

When k or $r = 0$, we refer to the resonant frequency f_0. We introduce the slow-varying envelope [30,36] to give

$$\widetilde{A}_{n\pm1} = A_{n\pm1} \exp(j2\pi t(\mp\Delta_{kr})). \tag{7}$$

Using (7) in (5) with (2) and (1), we have

$$I(f) = \sum_n \widetilde{I}_{n,\mathrm{Total}} \delta(f - nf_0), \tag{8}$$

where, as shown in Appendix A in (A7),

$$\widetilde{I}_{n,\mathrm{Total}} = -\sum_k \sum_r M\alpha_k \gamma_r \left(\widetilde{A}_{n+1} - 2A_n + \widetilde{A}_{n-1} \right) \exp(j2\pi tn\Delta_{r0}). \tag{9}$$

For each mode n, we extract the perturbed amplitude as

$$\widetilde{I}_n = \widetilde{I}_{n,\mathrm{Total}} - I_n, \tag{10}$$

with unperturbed amplitude as

$$I_n = -M(A_{n-1} - 2A_n + A_{n+1}). \tag{11}$$

2.2. Effect on Pulsetrains

To bring out the effect of coupling with multiple inputs, we separate $\widetilde{I}_n$ as

$$\widetilde{I}_n = \widetilde{I}_{n,R} + \widetilde{I}_{n,X}, \tag{12}$$

where $\widetilde{I}_{n,X}$ is generated by coupling between the inputs and $\widetilde{I}_{n,R}$ consists of terms generated by the same input as previously noted [30]

$$\widetilde{I}_{n,R} = I_n \sum_r \gamma_r [\exp(j2\pi tn\Delta_{r0}) - 1]. \tag{13}$$

Algebraic simplification for $\widetilde{I}_{n,X}$ (see (A8)–(A12) in Appendix B) gives

$$\widetilde{I}_{n,X} = -j4M \sum_r \sum_{k>r} \Gamma_{nrk} D_{nrk} S_{nrk}. \tag{14}$$

where, for any (r,k), we have generalized the definitions of β, Γ_n and Φ_n as

$$\beta_{rk} = \gamma_r \alpha_k = \gamma_k \alpha_r, \tag{15}$$

$$\Phi_{nrk} = \tan^{-1}\left[\frac{A_{n+1} - A_{n-1}}{A_{n+1} + A_{n-1}} \tan\left(2\pi tn\frac{\Delta_{rk}}{2}\right)\right], \tag{16}$$

$$\Gamma_{nrk} = \beta_{rk}\sqrt{A_{n+1}^2 + A_{n-1}^2 + 2A_{n-1}A_{n+1}\cos(2\pi tn\Delta_{rk})}, \tag{17}$$

$$D_{nrk} = \sin\left(2\pi t\frac{\Delta_{rk}}{2}\right)\cos\left(2\pi t\frac{\Delta_{rk}}{2} + \Phi_{nrk}\right) \tag{18}$$

$$\text{and } S_{nrk} = \exp\left(j2\pi tn\frac{\Delta_{r0} + \Delta_{k0}}{2}\right) \tag{19}$$

From (7), we note that, when $r = k$, $\widetilde{I}_{n,X} = 0$.

2.3. Phase Locking Condition and Locking Limits

Now, we look at the limits to injection locking for the phase condition found from the imaginary parts of Equations (3) and (12)

$$\sum_r \gamma_r \frac{2Qf_C}{\nu_s f_0} n\Delta_{r0} A_n = I_n \sum_r \gamma_r \sin(2\pi tn\Delta_{r0}) - 4M \sum_r \sum_{k>r} \Gamma_{nrk} D_{nrk} Re(S_{nrk}). \tag{20}$$

Let us understand qualitatively some of the effects in the phase relationship being captured by the terms in (20). The frequencies $n\Delta_{r0}$, $n\Delta_{k0}$ and $n\frac{\Delta_{r0} + \Delta_{k0}}{2}$, which appear in the terms Γ_{nrk} and $\frac{\Delta_{rk}}{2}$, have similar magnitudes. These frequencies are n times the beat frequency, $\frac{\Delta_{rk}}{2}$. When either the sin () or cos () terms go to zero in (20), the coupling has no influence on the injection range, leading to a time-varying phase locking relationship repeating with the beat frequency in time domain.

For injection locking between all frequencies r, the maximum frequencies $\Delta_{r0,\max}$ such that all the modes are still locked are possible when all the sinusoidal terms in (20) have unity magnitude to maximize the LHS; i.e.,

$$\sum_r \gamma_r \frac{2Qf_C}{\nu_s f_0} n\Delta_{r0,\max} A_n = I_n + 4M \sum_r \sum_{k>r} \Gamma_{nrk}. \tag{21}$$

Retaining only the frequency terms in the LHS of (21), we obtain the combined maximum locking frequency for all the inputs as

$$\sum_r \gamma_r \Delta_{r0,\max} = \frac{\nu_s f_0}{2Qf_C}\left(\frac{I_n}{A_n}\frac{1}{n} + \frac{4M\sum_r \sum_{k>r} \Gamma_{nrk}}{A_n}\frac{1}{n}\right). \tag{22}$$

Next, we look at the special case for two inputs where $r = \{1,2\}$ with $\Gamma_n = \Gamma_{n12}$ in (22)

$$\sum_{r=\{1,2\}} \gamma_r \Delta_{r0,\max} = \underbrace{\left(\frac{\nu_s f_0}{2Qf_C}\frac{I_n}{A_n}\right)\frac{1}{n}}_{\frac{R_n}{2}} + \underbrace{\left(\frac{\nu_s f_0}{2Qf_C}\frac{4M\Gamma_n}{A_n}\right)\frac{1}{n}}_{\frac{X_n}{2}}. \tag{23}$$

For two inputs, we have only one cross term evident. The first term in the RHS of (23) is the same as for perturbed AMLL applicable for single input given by $R_n = \frac{\nu_s f_0}{Qf_C}\left(\frac{I_n}{nA_n}\right)$ [30]. The increase in the total injection range is due to the second term in the RHS, which arises due to coupling between the modes given by

$$X_n = \frac{\nu_s f_0}{Qf_C}\left(\frac{4M\Gamma_n}{nA_n}\right). \tag{24}$$

For the two RF input cases, we obtain the total range by combining R_n with (24), $R_n + X_n$. We can therefore see that we can recover lock loss beyond the range R_n by using additional signals. To the author's knowledge, we are the first to demonstrate an increase in the effective locking range of laser modes without altering the physical parameters of the cavity. This extension of the effective locking range is as shown in (23).

3. Experiments and Results

An experiment with two sinusoidal inputs was set up to recover lock loss in higher modes as shown in Figure 1b. A function generator (AFG3252,Tektronix) was used to generate two RF inputs with constant phase relationship at frequencies $f_1 = f_0 + \Delta_{10}$ and $f_2 = f_0 + \Delta_{20}$. The inputs had detuning Δ_{10} and Δ_{20} from $f_0 = 26.69$ MHz, which were varied between 0 ± 13 kHz with 90 steps of size 289 Hz each. The two signals were added before feeding to the electro-optic modulator (EOM) of the AMLL. The AMLL construction is already described in our previous work and shown in Figure 1a [30,31]. The optical output of the laser was detected using a fast photodetector with 10 GHz bandwidth (DSC-R402AC, Discovery Semiconductors) and fed to an electrical spectrum analyzer (ESA). The spectrum was collected for each mode n around nf_0, with a resolution bandwidth of 500 Hz and span of 2 MHz. The peak amplitude (A_n) and frequency of the peak ($\widetilde{f_n}$) were recorded for each mode. The complete operation was automated using VISA standards in Python 2.7.

Figure 2 shows the spectra for modes $n = 19$ to 55, with and without a second input. In Figure 2a-left, only one input is present with $\Delta_{10} = -0.87$ kHz, $A = 4$ V-pp and $\alpha_1 = 1$. We can see a broad bell-shaped noise spectrum for the modes $n > 25$ without a distinguishable peak, indicating that they are unlocked. In Figure 2a-right, a second input signal with $\Delta_{20} = 1.16$ kHz, $\alpha_1 = 0.5$, $\alpha_2 = 0.5$ and $A = 4$ V-pp is added. Here, we see distinct peaks in the spectrum for modes $n > 25$, indicating relocking in the presence of the second input. To quantify the relocking effect of the second input, we measure the deviation $\left|\widetilde{\Delta}_n\right|$ as the difference between the expected peak frequency and the measured peak frequency, given by

$$\widetilde{\Delta}_n = \widetilde{f_n} - n(\alpha_1 f_1 + \alpha_2 f_2). \tag{25}$$

The deviation shown in Figure 2b indicates a lock loss in mode n when $\left|\widetilde{\Delta}_n\right| > 0$. For a single input, the modes $n > 25$ show unlocking. For $n = \{25 - 55\}$, $\left|\widetilde{\Delta}_n\right| \approx 0$ when the second signal is introduced, indicating a relocking of these modes. As a consequence of relocking, a gain of 14 dB is observed in the amplitude of the modes. We have thus demonstrated relocking of the cavity modes of an AMLL by use of two sinusoidal signals at the input and without altering the cavity's physical properties.

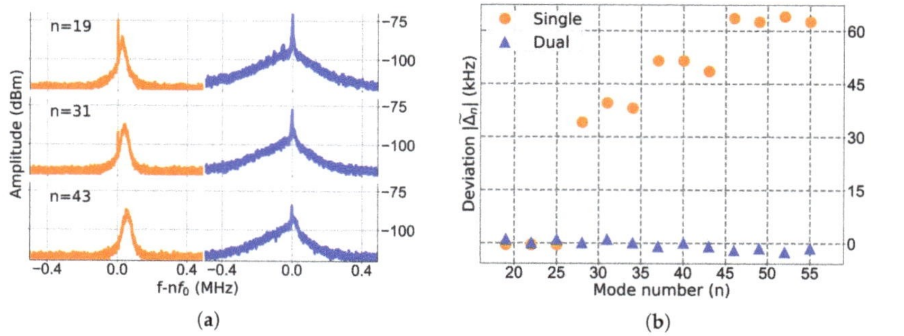

Figure 2. Comparison of two cases: single input with $\Delta_{10} = -0.87$ kHz (orange) and two inputs with $\Delta_{10} = -0.87$, $\Delta_{20} = 1.16$ kHz (blue). (**a**) Spectra for modes $n = 19, 31$ and 43. (**b**) Mode-wise deviation for modes $n = 19$ to 55.

Next, to find the limits on relocking frequencies detunings, (Δ_{10} and Δ_{20}) were varied by up to ± 13 kHz with step size $df = 289$ Hz, and deviation $\widetilde{\Delta}_n$ was recorded for modes $n = \{10, 30, 70\}$. Figure 3a shows the values of $\left|\widetilde{\Delta}_n\right|$ for mode $n = 10$. The white dots mark the edges of the locked regions such that $\widetilde{\Delta}_n < 10$ kHz, evaluated using a Laplacian filter with a threshold $\mathcal{T}$ set heuristically for a given mode [37]. For Δ_{10}, the maximum possible range is $R_{10} + X_{10}$ of 20.74 kHz within which relocking was successful, with the two inputs marked with dashed lines. The single input case where a range of $R_{10} = 14.34$ kHz was obtained for the same AMLL is also shown for comparison [30]. There is an increase of 6.4 kHz in the locking range for the 10th mode when two inputs are used. Similarly, an increase in range $R_n + X_n$ was observed when compared to R_n for $n = 30$ and 70, as shown in Figure 3b. To our knowledge, such an expansion of the locking range has been demonstrated in the work presented here for the first time.

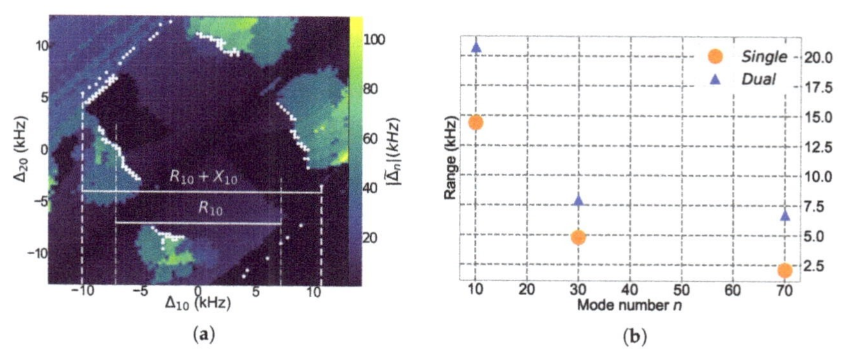

Figure 3. (**a**) Deviation for mode $n = 10$ with two injection signals ($\Delta_{10} = -0.87$, $\Delta_{20} = 1.16$ kHz). (**b**) Injection ranges for modes $n = 10, 30, 70$ for single- (R_n in orange) and two-input injection ($R_n + X_n$ in blue).

To look at the effect of relocking with a second input in the time domain, the time traces (using Textronix MDO3104 oscilloscope) and pulsewidth (using sampling oscilloscope (Lecroy SDA 100G)) after the fast photodetector were recorded. Figure 4 shows the pulsewidth (full-width half-maximum) for an average of 16 pulses triggered above the noise level. Figure 4a shows the pulsewidths for varying Δ_{10} and with and without the second input. For single input, the pulsewidth increases as $|\Delta_{10}|$ increases. Upon addition of a second input, one can observe a decrease in pulsewidth from approximately ~2 ns to ~205 ps, where the reduced pulsewidth matches the mode-locked AMLL pulsewidth for a single input with $\Delta_{10} = 0\,\text{kHz}$. Figure 4b shows the averaged pulse traces corresponding to points (i)–(iv) in Figure 4a, where $\Delta_{10} = -7, -2, -0.8$ and approximately 0 kHz, respectively. With two inputs, one can observe the reduced pulsewidth with a second RF input with $\Delta_{20} = 279, 359, 277$ and 211 kHz corresponding to $\Delta_{10} = -7, -2, -0.8$ and about 0 kHz, respectively. A reduction in pulsewidth is clearly visible. Note that pedestals in the traces could arise due to the averaging of pulse traces with a distribution of pulse shapes and widths between them, as predicted regarding the time-varying nature of the phase locking condition between the pulses in the pulse bunches (20). We believe that this is the first demonstration of a reduction in pulsewidth to mode-locked widths via relocking of higher modes with the use of two signals at the RF input in a fiber AMLL.

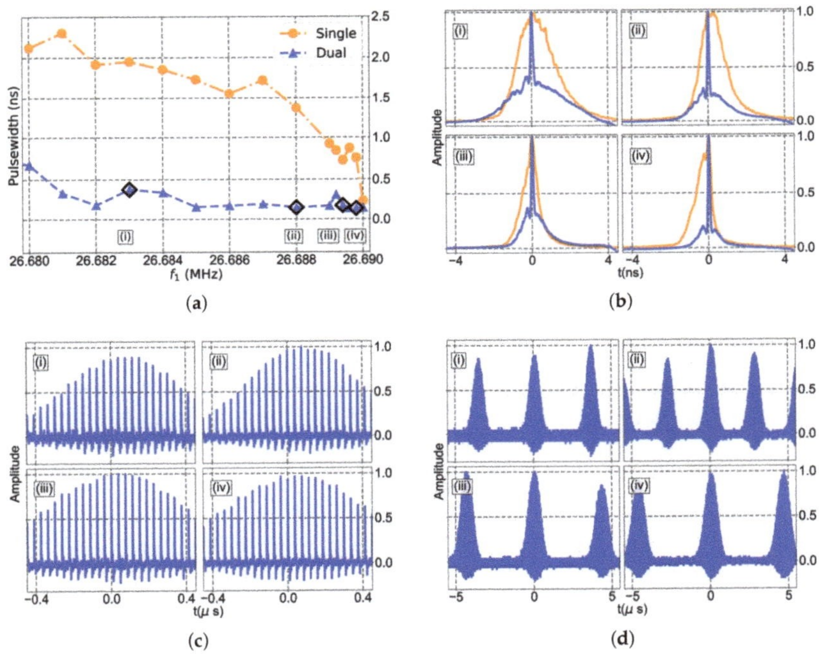

Figure 4. (**a**) Pulsewidth for different f_1 in the presence (blue) and absence (orange) of a second input. The points (i)–(iv) are taken as examples to demonstrate the effect a second input has on the pulses. (**b**) Pulses at points (i)–(iv) in (**a**) for single (orange) and two (blue) inputs. (**c**) ≈25 pulses in the pulsetrain for two inputs for cases (i)–(iv). (**d**) Pulsetrains for 11 µs for two inputs for cases (i)–(iv).

Figure 4c,d shows pulsetrains corresponding to the points (i)–(iv) with two inputs. Varying quality in pulses for about 25 pulses in each case is shown in Figure 4c. In Figure 4d, we have the pulsetrain for 11 μs for two inputs. We observe the occurrence of pulse bunches, consisting of many pulses separated by time periods of $3.77, 2.8$ and 4.73 µs corresponding to points (i), (ii) and (iv) with the corresponding $\Delta_{10} + \Delta_{20} = 265, 355$ and 211 kHz. This is in agreement with the theory presented in the previous section, where the slower time-varying nature in the phase condition is contained in the term D_{nrk} in (20). However, for point (iii), with $\Delta_{10} + \Delta_{20} = 275.2\,\text{kHz}$, the bunches themselves have a larger time period between them, which could be due to different operating regimes of operations and other non-linearities in the cavity.

4. Conclusions

In this paper, multiple RF input signals in an AMLL are shown as a novel method to extend the locking range of the cavity modes. We observe an increase in the number of locked modes and reduction in pulsewidths. A time-varying frequency model developed in this paper predicts locking of unlocked modes in a partially mode-locked laser when a second RF signal is used. The non-stationary model developed here also predicts the time-varying nature of the locking dynamics in the laser output with two inputs. Experimentally, it is demonstrated here that two sinusoidal RF inputs extend the locking range for 10th mode from $R_{10} = 14.34$ kHz by $X_{10} = 6$ kHz. The increased number of locked modes causes a narrowing of pulses in a partially locked laser from 2 ns to 205 ps.

Author Contributions: Conceptualization, S.K.; methodology, S.K.; software, S.K.; validation, S.K.; formal analysis, S.K.; investigation, S.K.; resources, A.P.; data curation, S.K.; writing—original draft preparation, S.K.; writing—review and editing, S.K. and A.P.; visualization, S.K.; supervision, A.P.; project administration, A.P.; funding acquisition, S.K. and A.P. All authors have read and agreed to the published version of the manuscript.

Funding: S. Krishnamoorthy thanks BERI for supporting her research and providing the opportunity to be associated with the project. She also thanks IPIC, Science Foundation Ireland (SFI) (Grant ID SFI/15/RP/2828 and SFI/22/RP-2TF/10293) and Tyndall National Institutes for providing sustenance during the manuscript preparation.

Institutional Review Board Statement: Not applicable.

Informed Consent Statement: Not applicable.

Data Availability Statement: Data underlying the results presented in this paper are not publicly available at this time but may be obtained from the authors upon reasonable request.

Acknowledgments: The authors thank Jayavel D., Yusuf Panbiharwala and Sathish for help with construction of the lasers. The authors thank Satyajit Mayor and Central Imaging & Flow Cytometry Facility (CIFF) at NCBS-TIFR, Bangalore, as well as the photonics@IITM group and Jitu lab for facilities, and Stefan Andersson-Engles, Julie Donelly, Simon Sorensen, Baptiste Jayet, Benjamin Lingnau, Michael Dillane, BioPhotonics@Tyndall, Aravind P. Anthur and Deepa Venkatish for useful suggestions. The authors wish to thank the anonymous reviewers for their valuable suggestions.

Conflicts of Interest: The authors declare no conflict of interest.

Appendix A. Derivation of $\widetilde{I_n}$

To observe the impact of this interaction on individual cavity mode, we recognize that each pulsetrain, $P_r(f)$, is composed of the contribution from each cavity mode n with amplitude A_n at frequency nf_r, such that one can write

$$P_r(f) = \gamma_r \sum_n A_n \delta(f - nf_r) \tag{A1}$$

where γ_r is the relative strength of the pulsetrain having repetition rate f_r such that $\sum_r \gamma_r = 1$. The total pulse in (1) can be expressed as

$$P(f) = \sum_{r \in \mathbb{N}} \gamma_r \sum_n A_n \delta(f - nf_r) \tag{A2}$$

Using (5) and (A2), we have

$$I(f) = -\sum_{k \in \mathbb{N}} M\alpha_k \sum_{r \in \mathbb{N}} \gamma_r \left[\sum_n A_n \delta(f - nf_r + f_k) - 2\sum_n A_n \delta(f - nf_r) + \sum_n A_n \delta(f - nf_r - f_k) \right]. \tag{A3}$$

To evaluate the effect of perturbation in the system described above, we note with indices $k, r \in \mathbb{N}$, the frequencies f_r and f_k are both detuned from f_0 such that $f_{(k \text{ or } r)} = f_0 + \Delta_{(k \text{ or } r)0}$ and $f_k = f_r + \Delta_{kr}$.

Using (A3) and regrouping the terms under a common summation in n, we have

$$I(f) = -M \sum_{k\in\mathbb{N}} \alpha_k \sum_{r\in\mathbb{N}} \gamma_r \sum_{n} \left(\widetilde{A}_{n+1} - 2A_n + \widetilde{A}_{n-1}\right)\delta(f - nf_r). \tag{A4}$$

Note that the effects of amplitude and phase change from the coupling are absorbed in the terms $\widetilde{A}_{n+1}$ and $\widetilde{A}_{n-1}$. Now, we can represent the total modulated amplitude, $I(f)$ in (A4), as a sum of $\widetilde{I}_{n,\text{Total}}$ for each mode n as

$$I(f) = \sum_{n} \widetilde{I}_{n,\text{Total}}\delta(f - nf_0), \tag{A5}$$

where we have the perturbed amplitude

$$\widetilde{I}_n = \widetilde{I}_{n,\text{Total}} - I_n, \tag{A6}$$

with I_n given by (11) and

$$\widetilde{I}_{n,\text{Total}} = -\sum_{k\in\mathbb{N}}\sum_{r\in\mathbb{N}} M\alpha_k\gamma_r \exp(j2\pi tn\Delta_{r0})\left(\widetilde{A}_{n+1} - 2\,A_n + \widetilde{A}_{n-1}\right). \tag{A7}$$

Appendix B. Two Input $\widetilde{I}_{n,X}$

To simplify the expression for the cross terms, we first look at a case with only two inputs. For two inputs, we expand $\widetilde{I}_{n,X}$ for $r = \{1,2\}$ and $k = \{1,2\}$ as

$$\widetilde{I}_{n,X} \triangleq -M2j\sin\left(2\pi t\frac{\Delta_{12}}{2}\right)\exp\left(j2\pi tn\frac{(\Delta_{10}+\Delta_{20})}{2}\right)$$
$$\left[\exp\left(j2\pi t\frac{\Delta_{12}}{2}\right)\Gamma_n\exp(j\Phi_n) + \exp\left(-j2\pi t\frac{\Delta_{12}}{2}\right)\Gamma_n\exp(-j\Phi_n)\right]. \tag{A8}$$

where we have noted that $\Delta_{12} = -\Delta_{21}$ and,

$$\Gamma_n\exp(j\Phi_n) = \gamma_1\alpha_2\exp\left(j2\pi tn\frac{\Delta_{12}}{2}\right)A_{n+1} + \gamma_2\alpha_1\exp\left(-j2\pi tn\frac{\Delta_{12}}{2}\right)A_{n-1}. \tag{A9}$$

Further, we note that the magnitude term Γ_n is real and positive, $0 \leq \alpha_1, \alpha_2 \leq 1$ and $0 \leq \gamma_1, \gamma_1 \leq 1$, which leads to the condition for cross terms as

$$\beta = \gamma_1\alpha_2 = \gamma_2\alpha_1. \tag{A10}$$

Using (A10) with (A9), we have

$$\Gamma_n = \beta\sqrt{\left(A_{n+1}^2 + A_{n-1}^2 + 2A_{n-1}A_{n+1}\cos\left(2\pi tn\Delta_{12}\right)\right)}. \tag{A11}$$

Next, from (A9) and (A10), we obtain the value for Φ_n as

$$\Phi_n = \tan^{-1}\left[\frac{A_{n+1} - A_{n-1}}{A_{n+1} + A_{n-1}}\tan\left(2\pi tn\frac{\Delta_{12}}{2}\right)\right] \tag{A12}$$

References

1. Cicerone, M.T.; Camp, C.H. Histological coherent Raman imaging: A prognostic review. *Analyst* **2018**, *143*, 33–59. [CrossRef]
2. Yang, B.; Fang, C.-Y.; Chang, H.-C.; Treussart, F.; Trebbia, J.-B.; Lounis, B. Polarization effects in lattice–STED microscopy. *Faraday Discuss.* **2015**, *184*, 37–49. [CrossRef] [PubMed]
3. Klar, T.A.; Jakobs, S.; Dyba, M.; Egner, A.; Hell, S.W. Fluorescence microscopy with diffraction resolution barrier broken by stimulated emission. *Proc. Natl. Acad. Sci. USA* **2000**, *97*, 8206–8210. [CrossRef] [PubMed]

4. Takasaki, K.T.; Ding, J.B.; Sabatini, B.L. Live-cell superresolution imaging by pulsed STED two-photon excitation microscopy. *Biophys. J.* **2013**, *104*, 770–777. [CrossRef] [PubMed]
5. Sidenstein, S.C.; D'Este, E.; Böhm, M.J.; Danzl, J.G.; Belov, V.N.; Hell, S.W. Multicolour multilevel STED nanoscopy of actin/spectrin organization at synapses. *Sci. Rep.* **2016**, *6*, 26725. [CrossRef]
6. Yao, X.S.; Davis, L.; Maleki, L. Coupled optoelectronic oscillators for generating both RF signal and optical pulses. *J. Light. Technol.* **2000**, *18*, 73–78. [CrossRef]
7. Yin, K.; Zhang, B.; Yang, W.; Chen, H.; Chen, S.; Hou, J. Flexible picosecond thulium-doped fiber laser using the active mode-locking technique. *Opt. Lett.* **2014**, *39*, 4259–4262. [CrossRef]
8. Xiao, K.; Jin, X.; Jin, X.; Yu, X.; Zhang, X.; Zheng, S.; Chi, H.; Feng, L.; Xu, M. Channelized amplification of RF signal based on actively mode locked fiber laser. *Opt. Commun.* **2018**, *421*, 46–49. [CrossRef]
9. Hjelme, D.R.; Mickelson, A.R. Theory of timing jitter in actively mode-locked lasers. *IEEE J. Quantum Electron.* **1992**, *28*, 1594–1606. [CrossRef]
10. Eichler, H.J.; Koltchanov, I.G.; Liu, B. Numerical study of the spiking instability caused by modulation frequency detuning in an actively mode-locked solid-state laser. *Appl. Phys. B* **1995**, *61*, 81–88. [CrossRef]
11. Wu, S.-Y.; Hsiang, W.-W.; Lai, Y. Synchronous-asynchronous laser mode-locking transition. *Phys. Rev. A* **2015**, *92*, 013848. [CrossRef]
12. Lee, J.; Lee, J.H. Experimental investigation of the cavity modulation frequency detuning effect in an active harmonically mode-locked fiber laser. *J. Opt. Soc. B Am. Opt. Phys.* **2013**, *30*, 1479–1485. [CrossRef]
13. Krishnamoorthy, S.; Mayor, S.; Prabhakar, A. Synchronization between two fixed cavity mode locked lasers. In Proceedings of the 5th International Conference on Photonics, Optics and Laser Technology, Porto, Portugal, 27 February–1 March 2017; Volume 1, pp. 273–282.
14. Krishnamoorthy, S.; Thiruthakkathevan, S.; Prabhakar, A. Active fibre mode-locked lasers in synchronization for STED microscopy. In *Optics, Photonics and Laser Technology 2017*; Springer: Berlin/Heidelberg, Germany, 2019; pp. 233–253.
15. Krishnamoorthy, S. Mode Unlocking and Relocking in a Detuned Actively Mode Locked Fiber Ring Laser. Ph.D. Thesis, Indian Institute of Technology Madras, Chennai, India, 2020.
16. Kuznetsov, A.; Kharenko, D.; Podivilov, E.; Babin, S. Fifty-ps raman fiber laser with hybrid active-passive mode locking. *Opt. Express* **2016**, *24*, 16280–16285. [CrossRef] [PubMed]
17. Kuznetsov, A.G.; Kablukov, S.I.; Timirtdinov, Y.A.; Babin, S.A. Actively mode locked raman fiber laser with multimode ld pumping. *Photonics* **2022**, *9*, 539. [CrossRef]
18. Yang, X.; Zhang, L.; Jiang, H.; Fan, T.; Feng, Y. Actively mode-locked raman fiber laser. *Opt. Express* **2015**, *23*, 19831–19836. [CrossRef]
19. Koliada, N.A.; Nyushkov, B.N.; Ivanenko, A.V.; Kobtsev, S.M.; Harper, P.; Turitsyn, S.K.; Denisov, V.I.; Pivtsov, V.S. Generation of dissipative solitons in an actively mode-locked ultralong fibre laser. *Quantum Electron.* **2013**, *43*, 5. [CrossRef]
20. Yao, J.; Yao, J.; Wang, Y.; Tjin, S.C.; Zhou, Y.; Lam, Y.L.; Liu, J.; Lu, C. Active mode locking of tunable multi-wavelength fiber ring laser. *Opt. Commun.* **2001**, *191*, 341–345. [CrossRef]
21. Lee, C.G.; Kim, Y.J.; Park, C.-S. Optical pulse shaping by cross-phase modulation in a harmonic mode-locked semiconductor fiber ring laser under large cavity detuning. *J. Light. Technol.* **2006**, *24*, 1237.
22. Nakazawa, M.; Yoshida, E. A 40 GHz 850 fs regeneratively FM mode-locked polarization-maintaining erbium fiber ring laser. *IEEE Photonics Technol. Lett.* **2000**, *12*, 1613–1615. [CrossRef]
23. Wise, F.; Lefrancois, S. Fiber Source of Synchronized Picosecond Pulses for Coherent Raman Microscopy and Other Applications. U.S. Patent 10,608,400, 31 March 2020.
24. Sato, K.; Ishii, H.; Kotaka, I.; Kondo, Y.; Yamamoto, M. Frequency range extension of actively mode-locked lasers integrated with electroabsorption modulators using chirped gratings. *IEEE J. Sel. Top. Quantum Electron.* **1997**, *3*, 250–255. [CrossRef]
25. York, R.A. Nonlinear analysis of phase relationships in quasi-optical oscillator arrays. *IEEE Trans. Microw. Theory Techn.* **1993**, *41*, 1799–1809. [CrossRef]
26. Yeung, M.S.; Strogatz, S.H. Time delay in the Kuramoto model of coupled oscillators. *Phys. Rev. Lett.* **1999**, *82*, 648. [CrossRef]
27. Habruseva, T.; Huyet, G.; Hegarty, S.P. Dynamics of quantum-dot mode-locked lasers with optical injection. *IEEE J. Sel. Top. Quantum Electron.* **2011**, *17*, 1272–1279. [CrossRef]
28. Shortiss, K.; Lingnau, B.; Dubois, F.; Kelleher, B.; Peters, F.H. Harmonic frequency locking and tuning of comb frequency spacing through optical injection, *Opt. Express* **2019**, *27*, 36976–36989. [CrossRef] [PubMed]
29. Lingnau, B.; Shortiss, K.; Dubois, F.; Peters, F.H.; Kelleher, B. Universal generation of devil's staircases near hopf bifurcations via modulated forcing of nonlinear systems. *Phys. Rev. E* **2020**, *102*, 030201. [CrossRef]
30. Krishnamoorthy, S.; Prabhakar, A. Mode unlocking characteristics of an RF detuned actively mode-locked fiber ring laser. *Opt. Commun.* **2019**, *431*, 39–44. [CrossRef]
31. Krishnamoorthy, S.; Mayor, S.; Prabhakar, A.Mode re-locking in an RF detuned actively mode-locked fiber ring laser. In Proceedings of the The European Conference on Lasers and Electro-Optics, Munich, Germany, 23–27 June 2019.
32. Adler, R. A study of locking phenomena in oscillators. *Proc. IRE* **1946**, *34*, 351–357. [CrossRef]
33. Strogatz, S.H. *Nonlinear Dynamics and Chaos: With Applications to Physics, Biology, Chemistry, and Engineering;* Westview Press: Boulder, CO, USA, 2014.

34. Razavi, B. A study of injection pulling and locking in oscillators. In Proceedings of the IEEE 2003 Custom Integrated Circuits Conference, San Jose, CA, USA, 24 September 2003; pp. 305–312.
35. Haus, H.A. A theory of forced mode locking. *IEEE J. Quant. Electron.* **1975**, *11*, 323–330. [CrossRef]
36. Tse, D.; Viswanath, P. *Fundamentals of Wireless Communication*; Cambridge University Press: Cambridge, UK, 2005.
37. Gonzalez, R.C.; Woods, R.E. *Digital Image Processing*; Pearson-Prentice-Hall Upper Saddle River: Saddle River, NJ, USA, 2002.

photonics

MDPI

Article

Frequency Comb Generation Based on Brillouin Random Lasing Oscillation and Four-Wave Mixing Assisted with Nonlinear Optical Loop Mirror

Yuxi Pang [1,2], Shaonian Ma [1,2], Qiang Ji [1,2], Xian Zhao [1,2], Yongfu Li [1,2], Zengguang Qin [2,3], Zhaojun Liu [2,3] and Yanping Xu [1,2,*]

1 Center for Optics Research and Engineering, Shandong University, Qingdao 266237, China
2 Key Laboratory of Laser and Infrared System of Ministry of Education, Shandong University, Qingdao 266237, China
3 School of Information Science and Engineering, Shandong University, Qingdao 266237, China
* Correspondence: yanpingxu@sdu.edu.cn

Abstract: A frequency comb generator (FCG) based on dual-cavity Brillouin random fiber lasing oscillation in the 1.5 μm telecon spectral window is established and experimentally demonstrated. In the half-open main cavity of the dual cavity, the stimulated Brillouin scattering in highly nonlinear fiber (HNLF) and Rayleigh scattering in single-mode fiber are employed to provide sufficient Brillouin gain and the randomly distributed feedback, respectively, for random mode resonance. The sub-cavity includes an Er-doped fiber amplifier to couple back and boost lower-order Stokes and anti-Stokes light for the cascade of stimulated Brillouin scattering to generate multiple higher-order Stokes and anti-Stokes light. Meanwhile, efficient four-wave mixing is stimulated in the HNLF-based main cavity, further enhancing the number and intensity of the resonant Stokes and anti-Stokes light. By taking advantages of the unique transmission characteristics of nonlinear optical loop mirrors, the power deviation between Stokes and anti-Stokes lines is further optimized with 17 orders of stable Stokes lines and 15 orders of stable anti-Stokes lines achieved within the 10 dB power deviation, with maximum optical signal-to-noise ratio (OSNR) of ~22 dB and ~17 dB and minimum OSNR of ~10 dB and ~7.5 dB for Stokes and anti-Stokes lines, respectively. In addition, the dynamic characteristics of the proposed FCG have been experimentally investigated. Such an FCG with fixed frequency spacing will find promising applications in fields of optical communication, microwave, optical sensing, etc.

Keywords: frequency comb; four-wave mixing; random fiber laser; Brillouin scattering; Rayleigh scattering; nonlinear optical loop mirror; dynamic characteristics

Citation: Pang, Y.; Ma, S.; Ji, Q.; Zhao, X.; Li, Y.; Qin, Z.; Liu, Z.; Xu, Y. Frequency Comb Generation Based on Brillouin Random Lasing Oscillation and Four-Wave Mixing Assisted with Nonlinear Optical Loop Mirror. *Photonics* **2023**, *10*, 296. https://doi.org/10.3390/photonics10030296

Received: 29 January 2023
Revised: 28 February 2023
Accepted: 7 March 2023
Published: 11 March 2023

1. Introduction

Frequency comb generators (FCGs) based on fiber-optic lasing oscillation have received intense attention and active research due to their promising applications in fiber-optic communication (e.g., fiber-optic wavelength division multiplexing systems) [1], fiber-optic sensing [2], microwave photonics [3], optical component characterization [4], optical instrumentation testing [5], and so on. Up to now, a number of FCGs realized by fiber-optic lasing oscillation based on different principles and structures have been proposed, mainly including the introduction of actively modulated frequency-shifting devices in the resonant cavity of fiber lasers [6–8] and the incorporation of the passively modulated narrow-bandwidth multi-channel filter [9–11] or intensity-dependent loss structures [12,13] in the resonant cavity of fiber lasers with broadband gain. However, the active modulation method requires electrical devices, which will undoubtedly introduce large insertion loss and break the all-fiber structure of FCGs. The frequency comb obtained by passive modulation methods is heavily dependent on the fabrication accuracy of the modulating device, making the FCG hard to generate frequency combs with narrow linewidth and fixed frequency spacing, limiting the widespread application of FCG.

In recent years, random fiber lasers (RFLs) with randomly distributed feedback provided by Rayleigh scattering in optical fibers has received a lot of attention due to their unique properties. It is a dramatic step up from 3D random lasers whereby the one-dimensional waveguide constraints are provided by the optical fibers. Recently, works on RFLs based on a variety of gain mechanisms have been gradually reported, mainly including rare-earth-ion doped fiber amplifiers [14–17], non-linear stimulated scattering processes in optical fibers (e.g., Brillouin and Raman gain) [18–21] and semiconductor optical amplifiers [22–24]. In addition, the performances of RFL have been continuously improved, such as lower threshold [25–27], higher output power [28–30], wavelength tunability [31–33], higher coherence [18,22,34], narrower linewidth [18,22,34,35], lower noise [36–38] and more stable output power [26,39,40], which enables a wide range of applications for RFLs in optical measurement [41,42], microwave photonics [43,44], optical fiber communication [45–47], optical fiber sensing [48–53], laser imaging [54,55] and other fields [56,57]. More importantly, the fiber laser gained by the stimulated Brillouin scattering (SBS) mechanism can achieve multi-wavelength laser output with a fixed wavelength spacing thanks to the cascade of SBS and the Doppler shift provided by the acoustic-induced dynamic grating in the optical fiber [58]. In particular, the Brillouin random fiber lasers (BRFLs) can also achieve the narrow-linewidth multi-wavelength laser output thanks to the narrow gain bandwidth of stimulated Brillouin scattering, which significantly enhances the potential value of this light source for application [59]. As a result, this multi-wavelength Brillouin random fiber laser (MW-BRFL) has attracted a great deal of attention in the generation of frequency combs and has been heavily investigated. In 2018, Liang Zhang et al. first proposed a dual-cavity MW-BRFL, which combines a half-open resonant cavity and a feedback cavity, and achieved a high-quality multi-wavelength laser output. Its optical signal-to-noise ratio (OSNR), polarization state, and the number of Stokes lines were optimized in subsequent works by using non-uniform fiber [59], polarization-maintaining fiber [60,61] and random fiber grating [62], respectively. However, only 14 maximum Stokes lines were eventually achieved due to the limitation of the relatively high laser threshold, and the anti-Stokes lines have not been fully boosted to be studied and exploited for FCG due to the absence of an efficient excitation mechanism. Fei Wang et al. achieved a half-open linear-cavity MW-BRFL with multiple Stokes and anti-Stokes lines through four-wave mixing (FWM) by means of a highly nonlinear fiber (HNLF). However, the associated laser performance has not been optimized and some of the key lasing properties have not been explored, which are essential to characterizing the application prospects of the generated lasing combs [5,63].

In this work, we present and experimentally investigate a dual-cavity FCG based on Brillouin random lasing oscillation. The Brillouin gain in the resonant cavity is provided by the HNLF with a high nonlinear coefficient ($\geq$10 $W^{-1}\cdot km^{-1}$ @1550 nm). Furthermore, the FWM effect in the HNLF is effectively excited under phase-matching conditions, which not only increases the number of Stokes lines but also boosts the power of the anti-Stokes lines to make them resonate in the main cavity. In addition, the nonlinear optical loop mirror (NOLM) with unique transmission spectrum is used to balance the resonant powers among Stokes and anti-Stokes lines of different orders to achieve a smaller power deviation. In the experiment, key lasing properties of the proposed FCG are investigated in detail, including the optical spectrum, linewidth, relative intensity noise (RIN), frequency noise (FN) and temporal characteristics.

2. Experimental Setup and Operation Principle

The experimental configuration of the proposed FCG based on Brillouin random fiber lasing oscillation is shown in Figure 1, in which the dual-cavity structure consists of a main cavity on the right and a sub-cavity on the left. The pump light is provided by a semiconductor laser at the wavelength of 1550 nm with a maximum output power of

10 dBm and a linewidth of ~15 kHz (LIGHTPROMOTECH, NLLD 0175-3-34-2). After passing through an optical isolator (ISO), the first optical coupler (OC1), an erbium-doped fiber amplifier (EDFA), a variable optical attenuator (VOA) and the first polarization controller (PC1), the finely tuned pump light is launched into a 390 m long HNLF through the first circulator (CIR1) from port 1 to port 2 in the main cavity. The ISO is used to prevent any reflected light from returning to the pump laser to avoid possible damage. The EDFA enables power amplification for the pump laser, and the VOA is used to finely tune the power of the amplified pump light to investigate the pump-power-dependent characteristics of the FCG. PC1 is used to adjust the polarization state of the pump light to achieve maximum Brillouin gain in the 390 m long HNLF.

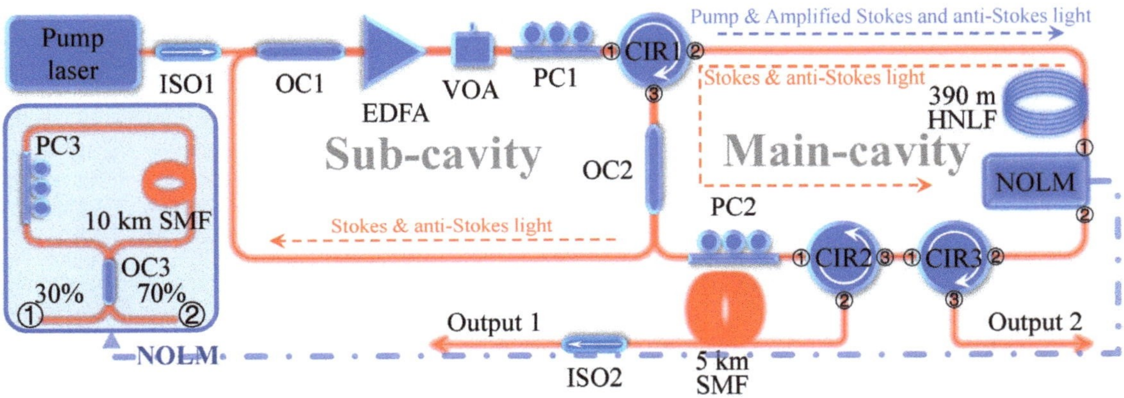

Figure 1. Experimental setup of the proposed FCG. (ISO: isolator, OC: optical coupler (OC1 and OC2: 2 × 2 50:50, OC3: 2 × 2 70:30), EDFA: erbium-doped fiber amplifier, VOA: variable optical attenuator, PC: polarization controller, CIR: circulator, HNLF: highly nonlinear fiber, SMF: single-mode fiber, NOLM: nonlinear optical loop mirror).

With the high non-linear coefficient ($\geq$10 $W^{-1}\cdot km^{-1}$ @1550 nm), the HNLF as a Brillouin gain medium provides sufficient gain for the random mode resonance in the main cavity. When pump light is launched into the HNLF, the spontaneous Brillouin scattering (SpBS) is first triggered by thermal noise and produces Stokes light propagating in the opposite direction to the pump light. Due to the Doppler shift effect, this Stokes light exhibits a frequency downshift of ~10 GHz compared to the pump light. The power of the Stokes light generated by SpBS increases as the pump power increases. When the pump power reaches the threshold of the SBS in the Brillouin gain medium, the Stokes light initiated by SpBS will be significantly amplified by the SBS process. The amplified backward-propagating Stokes light in the Brillouin gain medium passes sequentially through ports 2 and 3 of CIR1 and is divided equally by OC2 into two beams. One beam enters the sub-cavity for further amplification and feedback, and the other beam continues propagating in the main cavity. In the main cavity, this part of the Stokes light then enters the 5 km single-mode fiber (SMF) through the second CIR (CIR2) from port 1 to port 2. Due to the non-uniform distribution of the refractive index along the fiber axis in the core of SMF, a small portion of the Stokes light is scattered back into the main cavity by the numerous scattering centers accumulated along the SMF as randomly distributed feedback for the random mode resonance. The constructive interference between multiply scattered Stokes light from the scattering centers results in coherent random spikes. The random feedback provided by the SMF for the Stokes light is eventually emitted from port 3 of CIR2 and re-enters the Brillouin gain medium after passing through the NOLM and the port1 and port2 of CIR3 to be further amplified as the seed for the SBS process. As the Stokes light is amplified during its cyclic reciprocation in the main cavity, its power increases rapidly. When the effective gain provided by the SBS process in the main cavity can overcome the total loss, the amplified Brillouin Stokes light will resonate in the main cavity and emit at the end of the 5 km SMF as the laser output of the BRFL. Meanwhile, the other beam split by OC2, identical to the resonant Stokes light in the main cavity, is injected into the sub-cavity, where it is recombined with the output of the semiconductor pump laser via OC1 and input into the EDFA at the same time. After being amplified by the EDFA, the resonant Stokes light is re-injected into the main-cavity via CIR1 as the pump light to generate the higher-order Stokes lights.

In the 390 m HNLF, multi-order Stokes resonant lights that satisfy the phase-matching condition is prone to induce FWM, which can boost Stokes and anti-Stokes light power and produce more higher-order Stokes and anti-Stokes lines. To satisfy the phase-matching condition for FWM in HNLF, the phase mismatch factor κ ($\kappa = \Delta\beta + 2\gamma P$) should be made as close to zero as possible. Here, γP is the nonlinear phase shift caused by self-phase modulation and cross-phase modulation, γ represents the nonlinear coefficient of fiber, and P denotes the pump power. Therefore, γP is always greater than zero. $\Delta\beta$ is the linear phase shift caused by different propagation constants, and it is greater than zero in the normal dispersion region, so κ will be positively beyond the zero point in this dispersion region. At the zero-dispersion wavelength, $\Delta\beta$ equals zero, and κ still remains greater than zero with the positive value of γP. Only in the anomalous dispersion region, where $\Delta\beta$ is negative, can it offset part of the positive value of γP and make κ closer to zero, meeting the phase-matching condition for FWM. The zero-dispersion wavelength of the HNLF used in this experiment is located within a narrow wavelength range near 1521 nm. Pumping with a 1550 nm laser can satisfy the phase-matching condition for FWM. In the formula for the phase mismatch factor, $\kappa = \Delta\beta + 2\gamma P$, $\Delta\beta = -2\pi c/\lambda_0{}^2\ [D\ (\lambda_P - \lambda_S)^2]$. After substituting the parameters in the expression of κ using the values in Table 1 for calculation, we can get $\kappa = 0.00039143\ \mathrm{m}^{-1}$, which is very close to zero. It is worth noting that the forward-propagating multi-order Stokes light amplified by the EDFA is theoretically more susceptible to induce FWM due to its higher power than the backward-propagating Stokes light that has just been amplified or generated by the SBS process. Consequently, the frequency comb with multiple comb spikes and fixed frequency spacing can be efficiently generated at both output port 1 and output port 2.

Table 1. Experimental parameters for the calculation of κ.

Parameter Name	Value
Dispersion (D) @1550 nm	0.5753183 ps/nm·km
Pump wavelength (λ_p)	1550 nm
Signal wavelength (λ_s) @15th-Stokes	1551.14 nm
Zero-dispersion wavelength (λ_0)	1521.78962 nm
Nonlinear coefficient (γ)	10 $\mathrm{W}^{-1}\cdot\mathrm{km}^{-1}$
Pump power (P)	50 mW

The configuration of the NOLM is shown in the inset of Figure 1, which is constructed by a 70:30 OC3, a PC3 and a 10 km long SMF. PC3 was used for polarization biasing of the passing light in this loop to introduce a different phase difference. The 10 km SMF is utilized to provide the accumulation of nonlinear phase shift for the passing light. The transmission function of the NOLM can be written as

$$T = 1 - 2\alpha(1 - \alpha)\{1 + \cos[\theta + (1 - 2\alpha)\phi]\} \tag{1}$$

where α is the coupling ratio of the OC3, θ is the additional phase difference induced by the PC3, $\phi = 2\pi n_2 P_i L/\lambda A_{\mathrm{eff}}$ is the nonlinear phase shift, n_2 is the nonlinear refractive index coefficient of SMF, L is the length of loop, λ is the operating wavelength, A_{eff} is the effective fiber core area, and P_i is the input power for NOLM. Using Equation (1), the transmission characteristics of the NOLM with 75 km and 10 km long SMF were simulated for different

θ ($\theta = 0.5\pi$, 0.667π, π, 1.333π, 1.5π, i.e., different settings of PC3) as a function of input power, and the results are shown in Figure 2. The specific values used in the simulation are $n_2 = 3.2 \times 10^{-20}$ m^2/W, $A_{eff} = 50$ μm^2, $\alpha = 0.3$ and $\lambda = 1550$ nm. It can be seen in Figure 2, under specific PC3 setting, that the transmissivity of the NOLM can increase or decrease with increasing input power. When PC3 is set at a position where the transmissivity of the NOLM decreases as the input power, the NOLM can function as a power balancer, in which the higher power beam will experience greater loss than the lower power beam. According to the simulation results, the NOLM with 75 km-long delay fiber has a good power balance performance at this power level, and this performance will deteriorate if the delayed fiber is cut short. However, it is worth noting that the NOLM with long delay fiber is bound to cause large losses. So, to ensure efficient laser resonance, the delay fiber length of the NOLM in this case is set to 10 km and the corresponding simulation results of transmission characteristics are shown in Figure 2b. In addition, the coupling coefficient of the OC3 constituting the NOLM make an impact on the transmissivity of NOLM. As shown in Figure 3, the dependence of transmissivity on the input power of NOLMs with OCs of different coupling rates is analyzed. It can be seen from the figure that when the coupling coefficient of OC is 0.3 (i.e., coupling ratio of 70:30), the rate of change of the transmissivity of NOLM with the incident optical power is relatively large, regardless of the phase difference introduced by PC3. It is also observed from Figure 3 that the transmissivity of NOLM is completely insensitive to the incident optical power when the coupling rate of OC3 is set to 0.5. This is because the NOLM has to be established by using an OC with unequal split ratio so that the optical powers of the two light components propagating in the opposite directions within the loop are unequal, allowing the two lights to generate and accumulate a nonlinear phase shift difference in the delayed fiber, thus making the transmissivity of the NOLM dependent on the incident optical power.

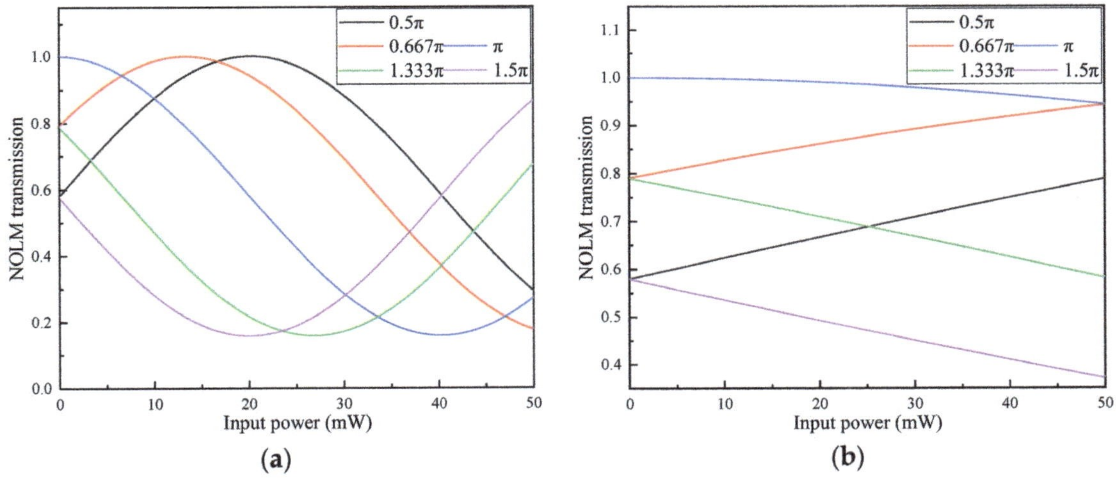

Figure 2. Simulation of the transmission characteristic of NOLM as a function of input power for different θ with (**a**) 75 km long and (**b**) 10 km long delay fiber, respectively.

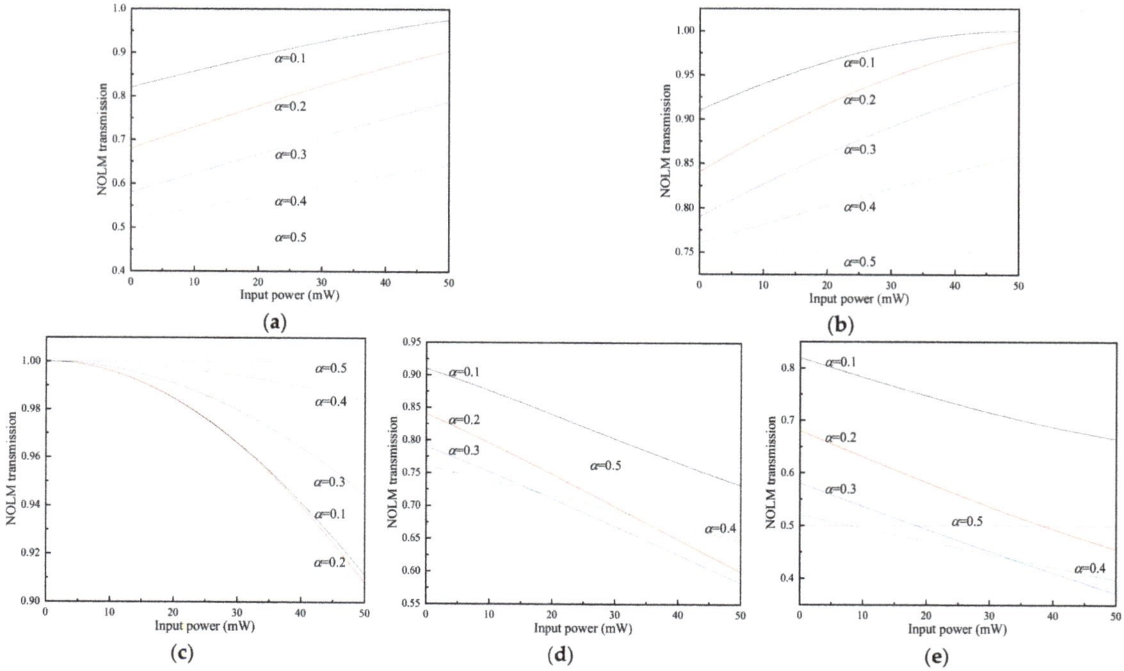

Figure 3. Simulation of the transmission characteristic of NOLM with different coupling ratios of OC3 as a function of input power for different θ of (**a**) 0.5π, (**b**) 0.667π, (**c**) π, (**d**) 1.333π and (**e**) 1.5π with 10 km long delay fiber.

3. Experimental Results and Analysis

3.1. Optical Spectrum

Firstly, the first-order Stokes light is resonated in the main cavity with the Brillouin gain and the effective randomly distributed feedback when the EDFA-amplified pump light was launched into the 390 m HNLF with above-threshold power. At the same time, a portion of the amplified Stokes light is guided into the sub-cavity by passing through OC2, where it is amplified by the EDFA and then sent into the main cavity as the pump light for the generation of Stokes lines of subsequent order. The Stokes light can resonate when the Brillouin gain in the main cavity overcomes the total roundtrip loss. As the output power of the EDFA increases, the number of Stokes lines generated in the main cavity gradually increases. Owing to the nature of cascaded SBS, the generated multi-order Stokes and anti-Stokes light is bound to have a power drop as the order increases, limited by the gain saturation of the EDFA. The cascaded SBS process will cease to generate higher-order Stokes light when the power of newly generated highest-order Stokes light does not reach the threshold for the SBS process to generate the next-order Stokes light.

In the experiment, the spectra of the random lasing emitted from output port 1 and output port 2 of the FCG are monitored by an optical spectrum analyzer (OSA) (YOKOGAWA, AQ6370D). With the pump laser output power of 8.65 mW and EDFA output power of 500 mW, the frequency comb with a fixed wavelength spacing of 0.076 nm were emitted at output port 1 and output port 2, and their optical spectra are shown in Figure 4. Figure 4a shows the spectrum of the forward-propagating frequency comb generated at the output port 2 within a wavelength range of 12 nm, and Figure 4b shows an enlarged view of this spectrum in a wavelength range of 2.2 nm. Up to 17 stable Stokes lines and 15 stable anti-Stokes lines with maximum OSNR of ~22 dB and ~17 dB and minimum OSNR of ~10 dB and ~7.5 dB were obtained within a 10 dB power deviation, respectively. Similarly, Figure 4c shows the spectrum of the backward-propagating frequency comb generated at the output port 1 within a wavelength range of 12 nm, and Figure 4d shows an enlarged view of this spectrum in a wavelength range of 2.2 nm. Comparing the spectra of the frequency comb generated at the two output ports, it can be seen that for both Stokes and anti-Stokes lines, the number of forward-propagating frequency comb lines measured at output port 2 is higher than that at output port 1 within a certain power deviation. Since the measurement of the spectrum is performed by the OSA scanning from short to long wavelength range over a period of time, the fluctuations within the spectrum reflect the wavelength instability of the light source to be measured. It is clear from the power fluctuations at the bottom of the scanned spectra that the stability of the forward-propagating frequency comb shows a higher level than that of the back-propagating frequency comb. The reason for this is that the forward-propagating Stokes light is justly amplified by the EDFA, propagates in the same direction as the amplified pump light and exhibits higher FWM efficiency, producing multiple idle lights with up-frequency-shift and down-frequency-shift and thus achieving a greater number of Stokes and anti-Stokes lines with higher stability. It is worth noting that the power difference (<5 dB) between first-order Stokes line and first-order anti-Stokes lines at output port 2 is significantly reduced compared to the power difference (~20 dB) at output port 1, which presents a better power flatness of the overall frequency comb. In addition, the minimum OSNRs of the forward-propagating and backward-propagating frequency comb within the power deviation are ~10 dB and ~28 dB, respectively, both of which exhibit relatively high capability of noise suppression, which is mainly due to the suppression effect of the random resonance in the main cavity on the spontaneous radiation generated by the EDFA. However, due to the fact that the forward-propagating Stokes and anti-Stokes light do not undergo random resonance processes to compress the generated noise after being amplified by the EDFA and before radiating at port 2, it thus exhibits a deterioration in OSNR.

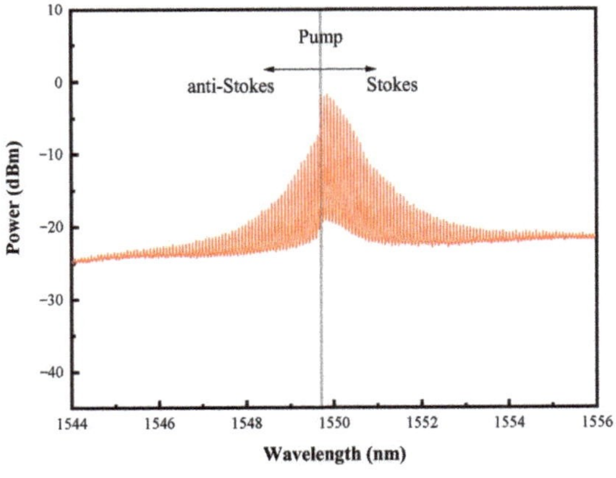

(a)

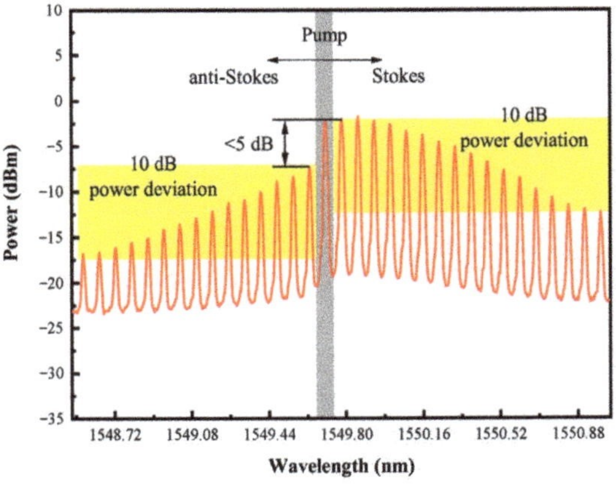

(b)

Figure 4. *Cont.*

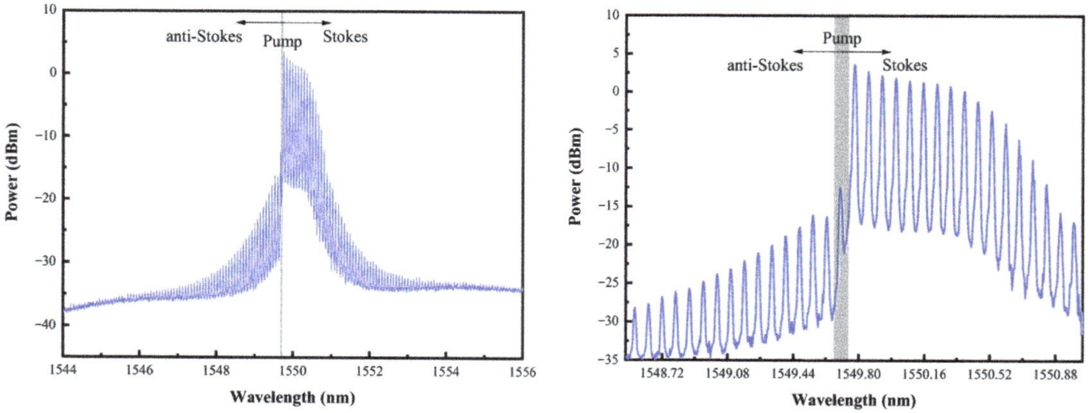

(c) (d)

Figure 4. Optical spectrum of (**a**) forward-propagating frequency comb in 12 nm wavelength span, (**b**) forward-propagating frequency comb in 2.2 nm wavelength span, (**c**) backward-propagating frequency comb in 12 nm wavelength span and (**d**) backward-propagating frequency comb in 2.2 nm wavelength span.

To investigate the influence of the EDFA output power on the generated frequency combs, the pump laser output power was set to 8.65 mW and the EDFA output power was gradually adjusted from 150 mW to 850 mW, while the frequency combs generated at the output port 2 and output port 1 were monitored by an OSA within a wavelength range of 2.6 nm; the measurement results are shown in Figure 5a,b, respectively. When the EDFA output power is 150 mW, the spectrum of frequency combs generated at output port 2 exhibits 18 Stokes lines and 16 anti-Stokes lines, and the spectrum of frequency combs generated at output port 1 shows 17 Stokes lines and 13 anti-Stokes lines. It is worth noting that the power flatness and stability of the frequency combs generated at output port 2 are significantly higher than that generated at output port 1. With increasing EDFA output power, the number and the respective intensity of Stokes and anti-Stokes lines of the frequency combs emitted at both output ports gradually increase, and meanwhile the power flatness of the frequency combs gradually improves with the assistance of NOLM.

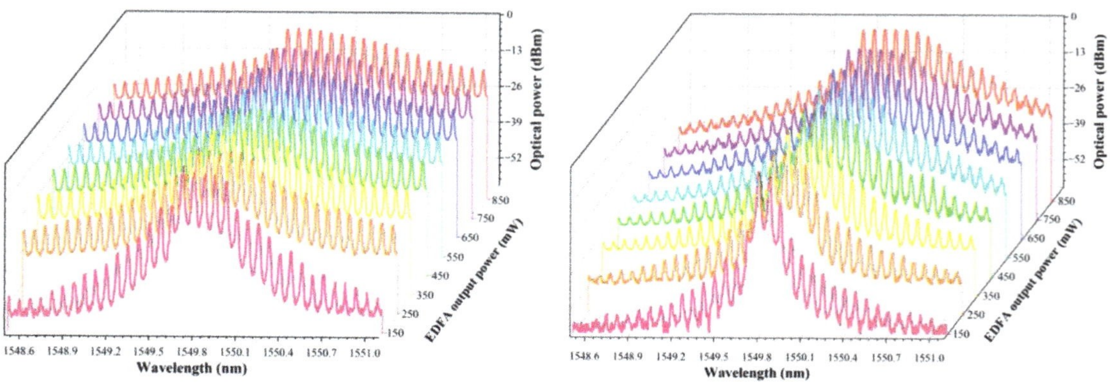

(a) (b)

Figure 5. Optical spectrum evolution of (**a**) forward-propagating and (**b**) backward-propagating frequency combs in 2.6 nm wavelength span with the increase of EDFA output power from 150 mW to 850 mW.

In the main cavity, both the FWM of Stokes light propagating in the forward and backward directions in the 390 m HNLF and the power balance provided by the NOLM are utilized to enhance the performance of the frequency combs based on Brillouin random fiber laser oscillation. Here, the performance enhancement effects of both mechanisms are further investigated. As shown in Figure 6a, the spectra of the frequency combs emitted at output port 2 and output port 1 with the EDFA output power of 850 mW and the pump laser output power of 8.65 mW are compared, and there is no NOLM addition at this time. It is obvious from the result that a larger number of Stokes and anti-Stokes lines in the forward-propagating frequency combs are effectively excited and their power flatness is significantly enhanced. This is because the backward-propagating Stokes light lacks the amplification provided by the EDFA, resulting in the more severe power decrease with the Stokes order due to the pump depletion effect. Consequently, the efficiency of FWM for higher orders of Brillouin Stokes and anti-Stokes light is significantly limited due to the low power of the high-order Stokes and anti-Stokes light, thus making it difficult for the anti-Stokes light to overcome the loss in the cavity and become resonant. In addition, in the backward direction which is opposite to the propagation direction of the pump light, only weak Rayleigh-scattered pump light joins the FWM process in that direction, which also limits the amplification for the backward-propagating anti-Stokes light through FWM. In contrast, the Stokes light in the forward-propagating frequency comb is amplified by the high-gain EDFA in the sub-cavity before it is launched into the HNLF. More importantly, the forward-propagating Stokes light propagates in the same direction with the high-power pump light, resulting in a much higher efficiency of the FWM than in the backward direction and facilitating the excitation and amplification of the forward-propagating anti-Stokes light. As shown in Figure 6b, after the NOLM was introduced into the main cavity, the flatness of the forward-propagating frequency combs was further improved by virtue of the unique input-power-dependent transmission characteristics of NOLM.

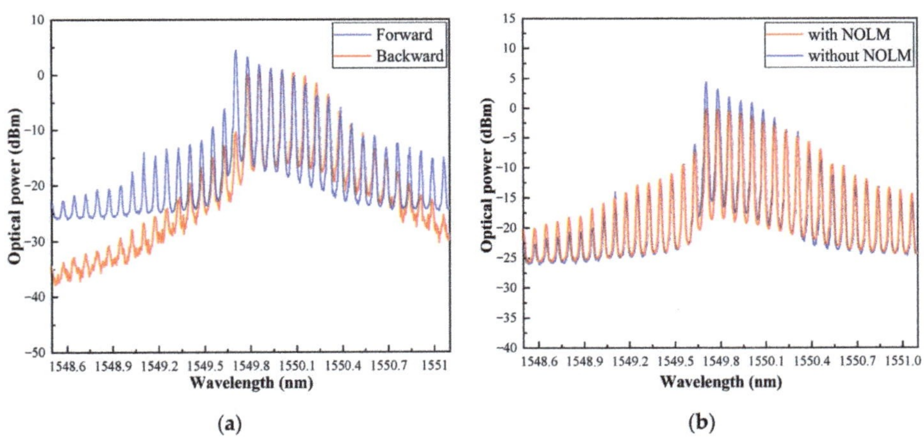

Figure 6. Optical spectrum comparison between (**a**) forward-propagating and backward-propagating frequency combs with EDFA output power of 900 mW; (**b**) forward-propagating frequency combs with or without NOLM.

3.2. *Linewidth*

Linewidth is an important parameter to demonstrate the coherence of the laser beam. To characterize the linewidth of each Stokes and anti-Stokes random radiation of the proposed FCG, a delayed self-heterodyne (DSH) measurement device was built. First, each order of Stokes and anti-Stokes random radiation in the generated frequency combs is filtered by a tunable bandpass filter (Alnair, BVF-300CL-SM-FA) with a set bandwidth of 3.7 GHz. The selected Stokes/anti-Stokes light is then split into two parts by a 50:50 OC;

one part is sent through an acousto-optic modulator (AOM) with an 80 MHz frequency downshift, and the other part is sent into a 100 km delayed line fiber. The two parts of the selected Stokes/anti-Stokes light with an 80 MHz frequency difference are then recombined at another 50:50 OC and form a beat light with a central frequency of 80 MHz. The generated beat light is detected by a photodetector (PD) (Thorlabs, PDB450) with a bandwidth of 350 MHz, the electrical spectrum of the beat light is analyzed by an electrical spectrum analyzer (ESA) and finally the spectral properties of selected Stokes/anti-Stokes light in the frequency domain is characterized.

The measured electrical spectrum of the beat signal of the Stokes lines from first-order to seventh-order is shown in Figure 7. Figure 7a shows an electrical spectrum comparison between the beat signal of the pump light and the Stokes light, from which it can be seen that the linewidth of the Stokes light is significantly narrowed compared to the pump light thanks to the SBS process. The 20 dB linewidths of the pump light and the first-order to seventh-order Stokes light were measured as 342.68, 8.687, 9.462, 8.823, 8.983, 13.475, 12.494 and 14.022 kHz, respectively. Correspondingly, the 3 dB linewidths of the pump light and Stokes lines were obtained by dividing the 20 dB linewidth by $2 \times (99)^{1/2}$, which are 17.22, 0.4365, 0.4755, 0.4434, 0.4514, 0.6771, 0.628 and 0.7046 kHz, respectively. All the measured Stokes lines in the generated frequency combs exhibit sub-kHz linewidths, indicating that the coherence of Stokes light is substantially enhanced as it is being resonated in the half-open cavity.

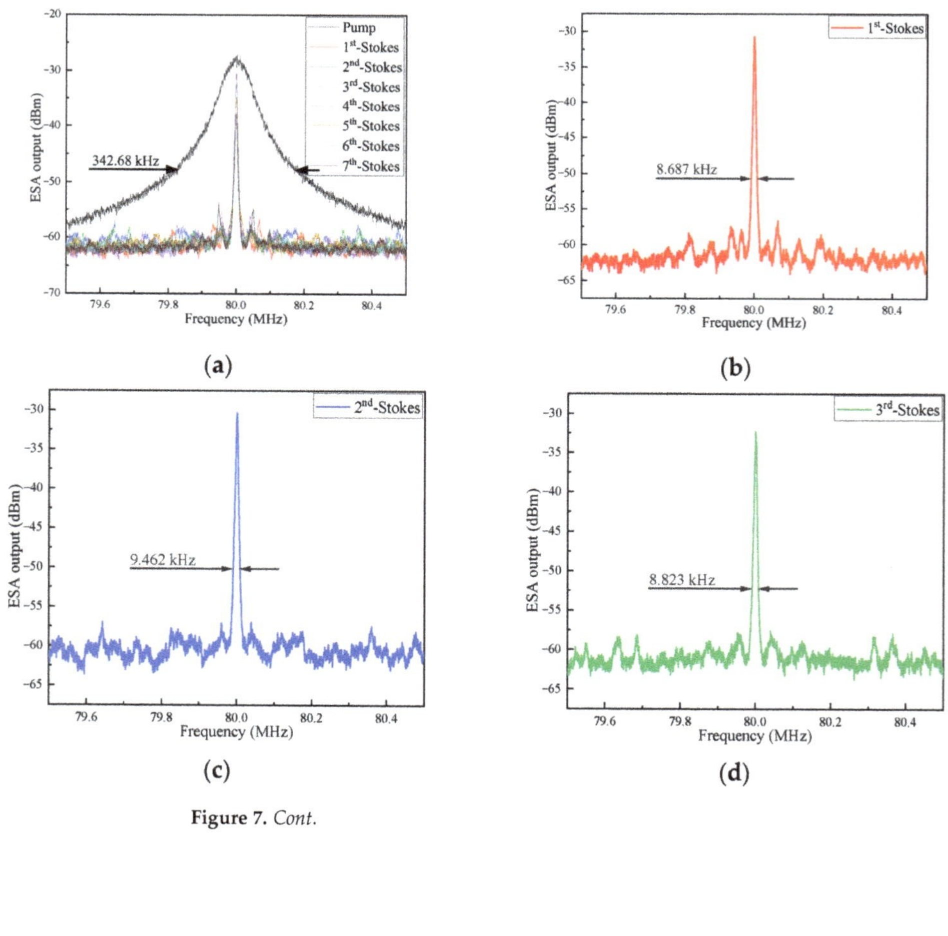

Figure 7. *Cont.*

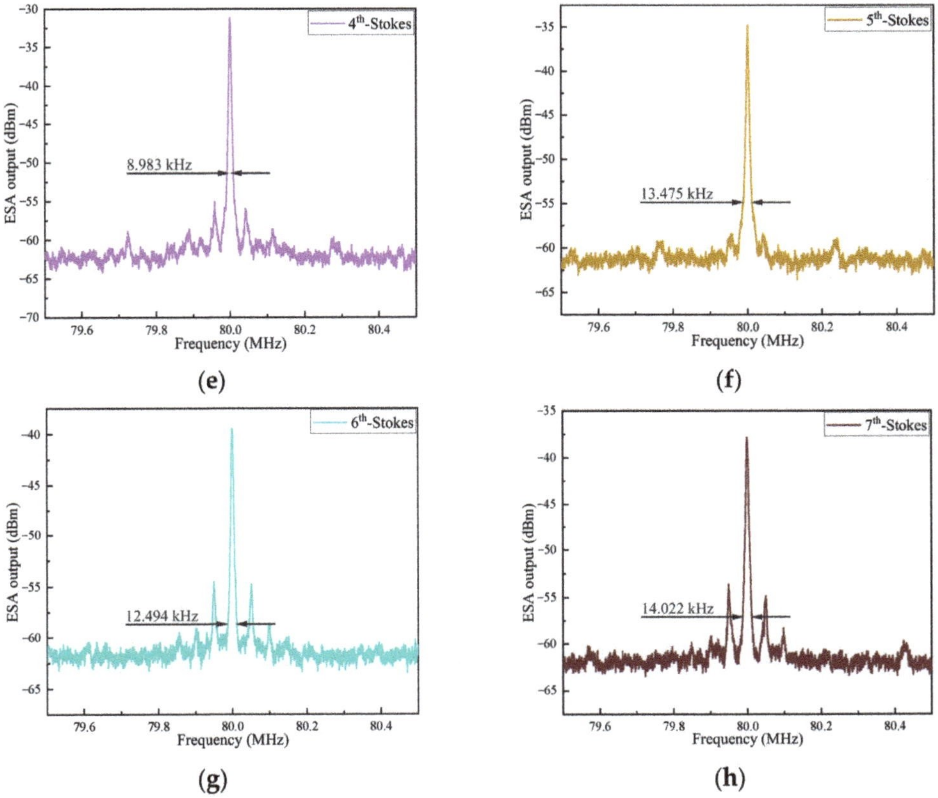

Figure 7. (**a**) DSH beating RF spectra of the pump laser and the 1st~7th order Stokes lines of the proposed FCG for comparison; (**b–h**) DSH beating RF spectra of the 1st~7th order Stokes lines of the proposed FCG, respectively.

In addition, the electrical spectra of the anti-Stokes lines from first-order to fourth-order in the generated frequency combs have also been experimentally investigated for the first time, and the measurement results are shown in Figure 8. The 20 dB linewidths of the first-order to fourth-order anti-Stokes light were measured as 54.948, 72.731, 124.943 and 171.911 kHz, respectively. Correspondingly, the 3 dB linewidths of the pump light and the first-order to fourth-order anti-Stokes lines were obtained by dividing the 20 dB linewidth by $2 \times (99)^{1/2}$, which are 2.76, 3.65, 6.28 and 8.64 kHz, respectively. From the electrical spectrum comparison of the four orders of anti-Stokes and pump light as shown in Figure 8a, it is clear that the linewidth of the anti-Stokes line is much smaller than that of the pump light, mainly due to the linewidth-narrowing effect during the SBS process. As can be seen in Figure 8f, the linewidth of the first-order anti-Stokes line is wider than that of the first-order Stokes line, and this phenomenon is also present for Stokes and anti-Stokes lines of other orders, which is due to the stronger dependence of the generation of anti-Stokes light on the FWM process than that of Stokes light, making the coherence of anti-Stokes lines slightly worse than that of Stokes lines. However, although anti-Stokes lines exhibit slight linewidth broadening, the linewidth of kHz-level still proves that the anti-Stokes lines generated by the proposed FCG are highly coherent, which can provide the frequency combs required for a variety of application areas.

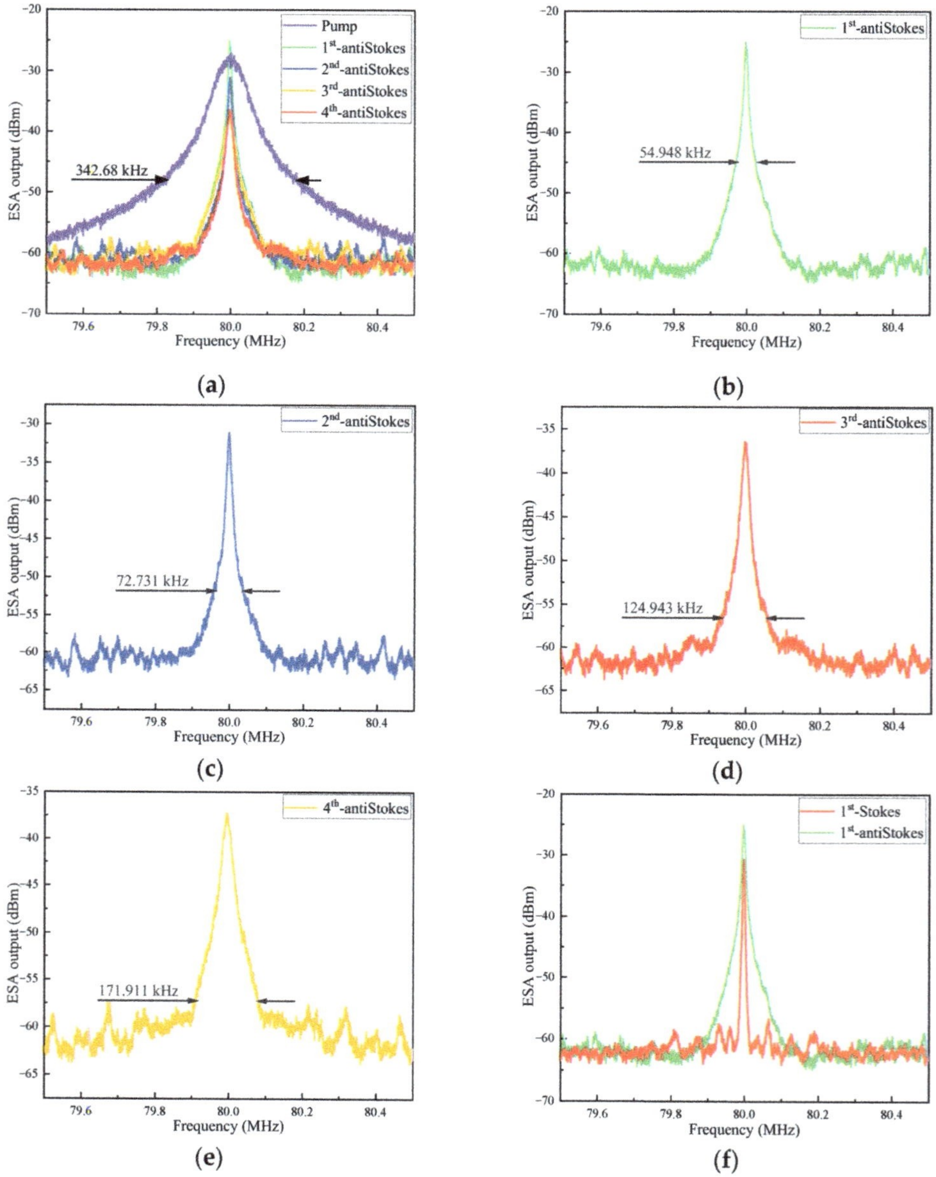

Figure 8. DSH beating RF spectra of (**a**) the pump laser and the 1st~4th order anti-Stokes lines of the proposed FCG for comparation; (**b**–**e**) DSH beating RF spectra of the 1st~4th order anti-Stokes lines of the proposed FCG; (**f**) comparison of the 1st-order-Stokes line and the 1st-order-anti-Stokes line.

3.3. Relative Intensity Noise

The relative intensity noise (RIN) of the Stokes and anti-Stokes light of the generated frequency combs was measured, which was achieved by first recording the random laser emission and then calculating the power fluctuation normalized to the average power level. Similar to the linewidth measurement, the Stokes/anti-Stokes lines of the generated frequency combs were firstly selected using a filter with a 3.7 GHz bandwidth. In the experiments, the RINs of the Stokes light from first order to seventh order and anti-Stokes light from first order to fourth order were measured and compared with commercial semiconductor lasers. Firstly, the power of the light to be measured was adjusted to the same magnitude of 1 mW by a VOA and the output power of the commercial laser was also set to 1 mW in order to have a fair comparison. The time domain power traces of the light to be measured were then obtained by using a PD (Thorlabs, PDB435) with a bandwidth of 150 MHz and recorded by an oscilloscope (TELEDY LECROY, HDO6104). After data processing, the RIN of Stokes light for the first seven orders and anti-Stokes light for the first four orders as well as that of the semiconductor laser were obtained, as shown in Figure 9.

It can be seen from Figure 9 that the RIN of each selected Stokes line and the anti-Stokes line is higher than that of the commercial semiconductor laser, especially in the low-frequency region of <1 kHz due to the effects of transient temperature variation and mechanical vibrations in the environment to the resonant cavity. Furthermore, the RIN of both the Stokes and anti-Stokes lines increases with their order due to the intensity fluctuations of the pump light as well as low-order Stokes and anti-Stokes light that are undoubtedly transferred to the generated higher-order Stokes and anti-Stokes light during the cascaded SBS process, resulting in higher RIN of subsequent order Stokes and anti-Stokes lines. It is worth noting that, as shown in Figure 9b, the RIN deterioration of anti-Stokes light with increasing order is more severe than that of Stokes light, due to the fact that most of the anti-Stokes lines originate from unstable FWM with lower coupling efficiency, leading to the introduction of more noise.

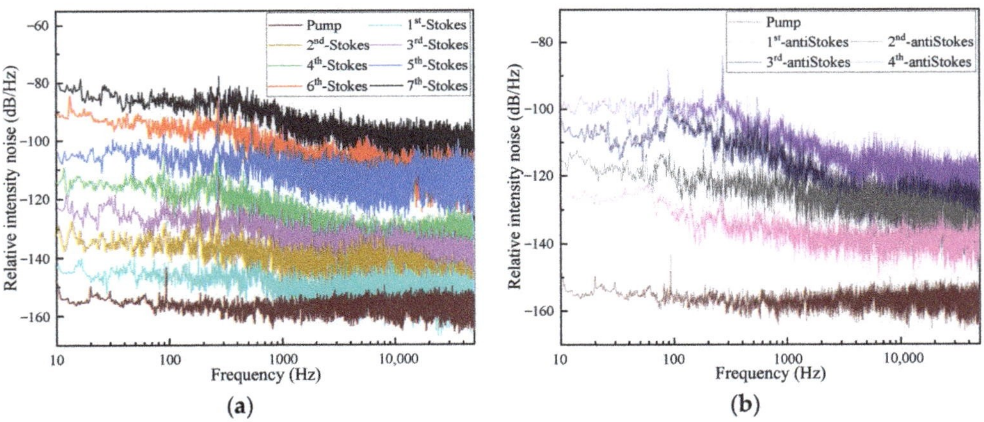

Figure 9. Relative intensity noise of (**a**) the Stokes random emissions of the first seven orders and the pump laser and (**b**) the anti-Stokes random emissions of the first four orders and the pump laser.

3.4. Frequency Noise

The frequency noise of selected Stokes and anti-Stokes lines is also investigated and compared with the commercial semiconductor laser. The measurement setup is a 3 × 3 unbalanced Michelson fiber interferometer. In this measurement system, a 2 km SMF was inserted into one arm as the delay line, and two Faraday rotation mirrors (FRM) at one side are used to reflect the light and eliminate the additional optical path difference due to the birefringence in the fiber. The output interference signal is detected at another side by two PDs and read by an oscilloscope. The frequency noise was then demodulated by implementing the differential cross-multiplication algorithm on the data collected by the two PDs at a 10 kHz measuring bandwidth [37]. Finally, the measured frequency noise spectra are plotted as shown in Figure 10.

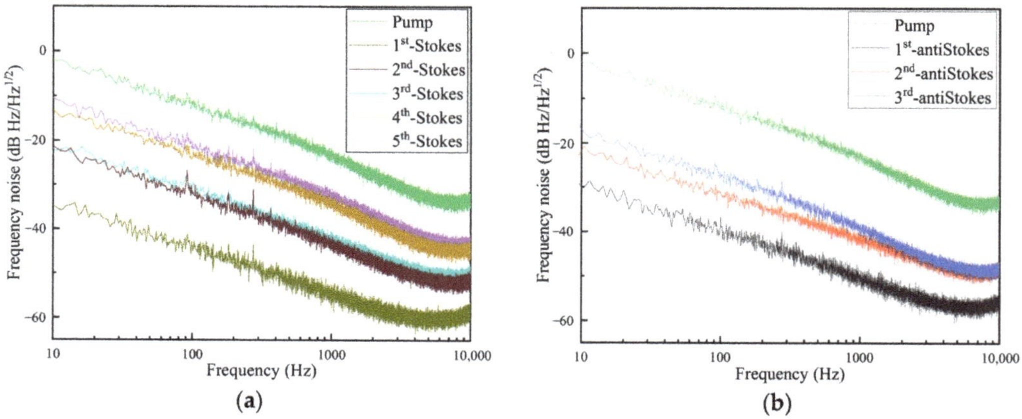

Figure 10. Frequency noise of (**a**) Stokes light of the first five orders and the pump laser and (**b**) the anti-Stokes light of the first three orders and the pump laser.

From Figure 10, it is clear that the frequency noise of both Stokes and anti-Stokes light is lower than that of commercial semiconductor lasers over the entire frequency range, which is mainly attributed to the suppression of the frequency noise in the half-open main cavity. Frequency noise suppression is mainly thanks to the suppression of the nominal $1/f$ noise from the thermal fluctuation that intrinsically exists in conventional cavity lasers because of the long cavity length and randomly distributed feedback from the Rayleigh scattering centers in the half-open main cavity. Furthermore, in agreement with the RIN measurement results, the frequency noise of both Stokes and anti-Stokes lines increases with the order number, which is caused by the transfer of noise from the pump light to the subsequently generated Stokes and anti-Stokes light during the cascade SBS process.

3.5. Temporal Characteristics

The temporal characteristics of Stokes lines in the generated frequency combs and its corresponding influence factors were experimentally investigated. Firstly, the pump power input to the EDFA was fixed at 3 mW and the EDFA output power was set at 350 mW. The temporal trace of the 1st-order Stokes lines selected by a filter with 3.7 GHz bandwidth was recorded by a PD and digitized by an oscilloscope. Then, the output power of the EDFA is gradually increased and the corresponding temporal trace of the 1st-order Stokes emission is shown in Figure 11. For fairness, the power of the first-order Stokes line to be measured generated with any EDFA output power is kept strictly at 1 mW by the VOA. From Figure 11, it can be seen that the first-order Stokes line exhibits a quasi-continuous laser emission when the EDFA output power is set at 350 mW. Subsequently, as the EDFA output power is increased, the temporal trace of the first-order Stokes line progressively exhibits more stochastic fluctuations and even sharp intensity fluctuation spikes, which are caused by the decrease in the proportion of the pump light at the input port of the EDFA as the proportion of Stokes and anti-Stokes light in the sub-cavity continues growing, absorbing most of the amplification provided by the EDFA and leading to a distinct reduction in the pump power amplification and a deteriorated SBS efficiency. When the output power of the EDFA reaches 450 mW, the first-order Stokes light exhibits severe stochastic intensity fluctuations, which indicates that the emission is only weakly resonated at this time.

In addition, in order to further demonstrate that the above-mentioned deterioration in the quality of the first-order Stokes light due to the decrease in the proportion of pump light in the EDFA input light, the relationship between the pump power and the state of the temporal trace of the first-order Stokes light has also been investigated and the results are shown in Figure 12. Firstly, the output power of the EDFA is set at 450 mW and the pump power is set at 3 mW, in which case the temporal trace of the first-order Stokes line exhibits severe stochastic intensity fluctuations as shown in Figure 12a. Subsequently, by gradually increasing the pump power from 3 mW to 5 mW, the time-domain intensity trace of the first-order Stokes line progressively stabilizes and the spikes of stochastic intensity fluctuation gradually decrease until they disappear, as shown in Figure 12a–e.

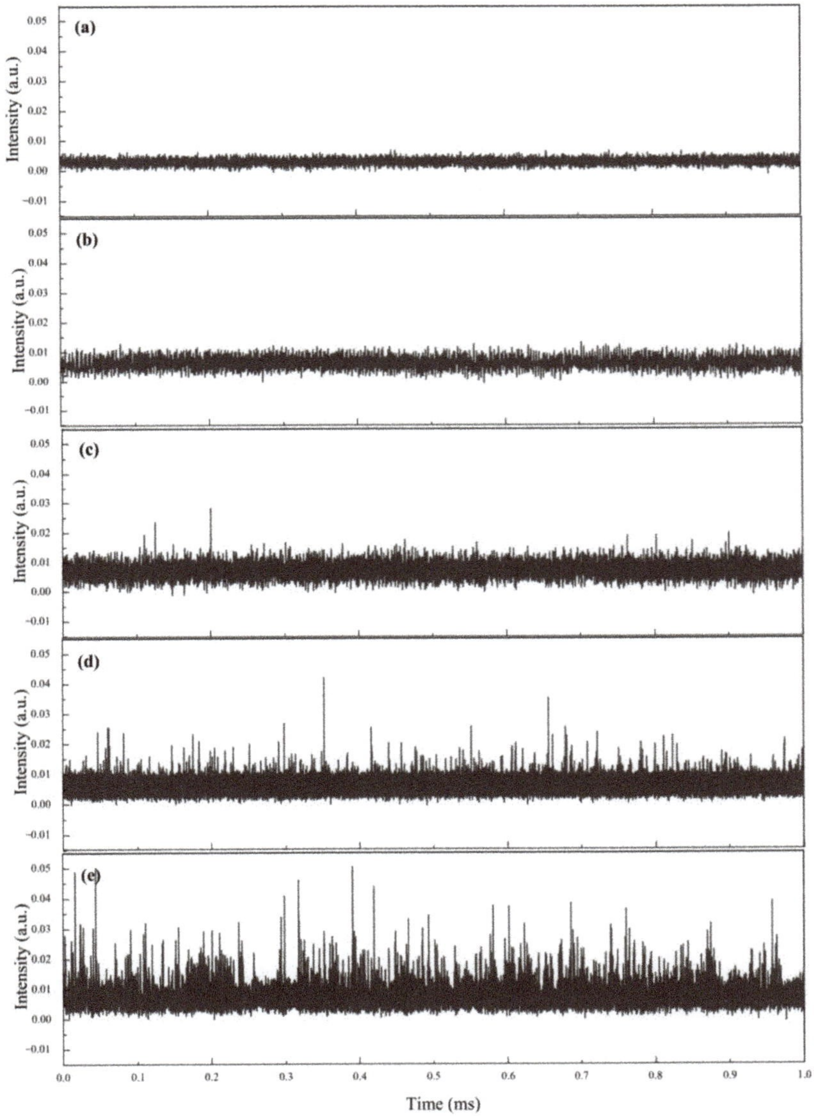

Figure 11. Temporal traces of 1st-order Stokes light in generated frequency combs at the EDFA output power of (**a**) 350 mW, (**b**) 375 mW, (**c**) 400 mW, (**d**) 425 mW and (**e**) 450 mW, and at the pump power of 3 mW.

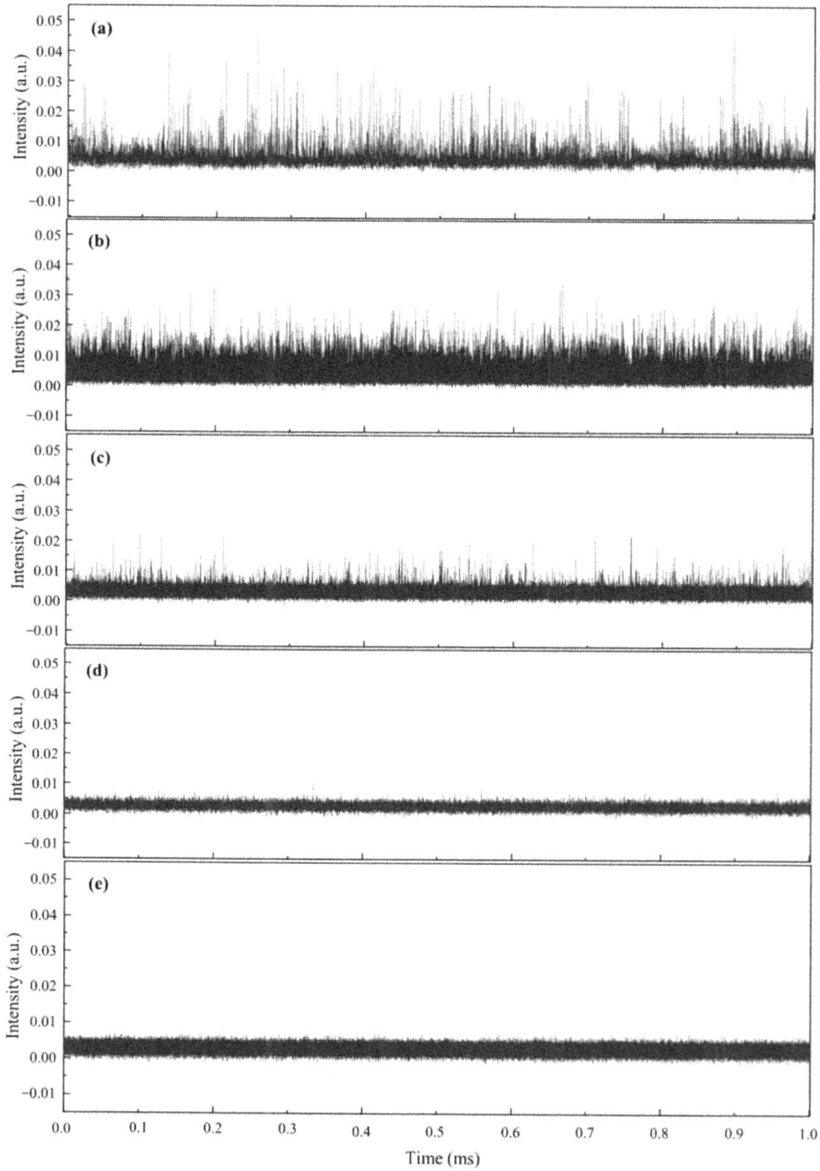

Figure 12. Temporal trace of first-order Stokes light in generated FCG at the pump power of (**a**) 3 mW, (**b**) 3.5 mW, (**c**) 4 mW, (**d**) 4.5 mW and (**e**) 5 mW, and at the EDFA output power of 450 mW.

The above experiment shows that the generation of frequency combs in the dual-cavity Brillouin random oscillation device requires attention to the adjustment of the ratio of the power of the pump light to the power of the feedback light in the sub-cavity, in order to achieve more Stokes lines for the generated frequency combs and better stability and coherence of each order of Stokes light.

The noise of different-order Stokes and anti-Stokes lines of the generated frequency combs has been investigated above, revealing the noise transfer characteristics present in this Brillouin-scattering-based FCG. Therefore, in order to further investigate the dynamic properties of the Stokes and anti-Stokes lines of different orders, the temporal traces of different-order Stokes and anti-Stokes lines were studied in detail. Figures 13 and 14 show the temporal traces of Stokes light of order one to seven and anti-Stokes light of order one to four filtered out by using a filter with 3.7 GHz bandwidth at the pump power of 8.65 mW and EDFA output power of 500 mW, respectively.

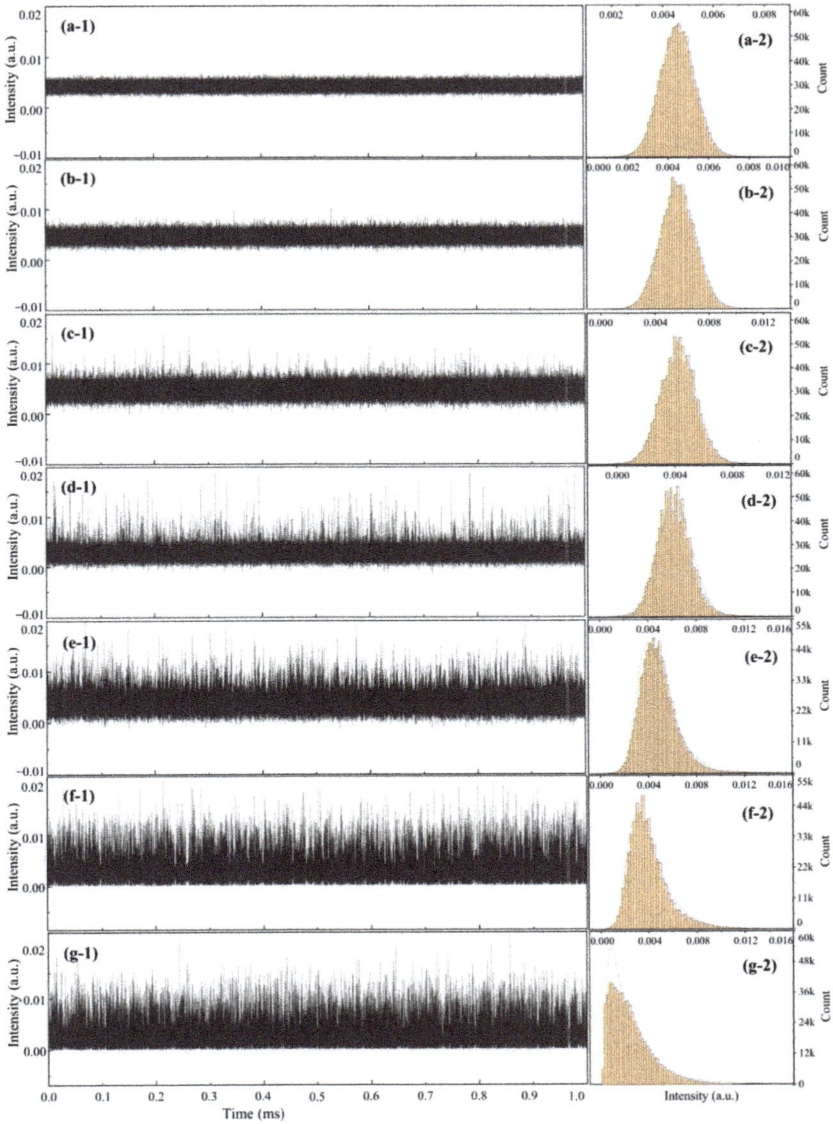

Figure 13. (**a-1–g-1**) Temporal trace and (**a-2–g-2**) intensity distribution of 1st-order~7th-order Stokes light in generated FCG.

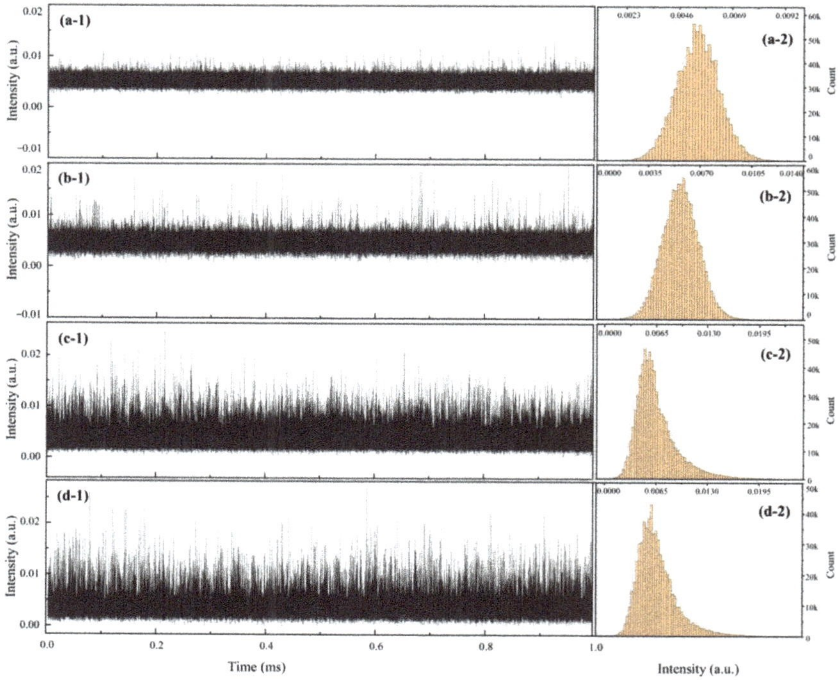

Figure 14. (**a-1**–**d-1**) Temporal trace and (**a-2**–**d-2**) intensity distribution of 1st-order~4th-order anti-Stokes lines in the proposed FCG.

As can be seen in Figure 13, the temporal trace of the first-order Stokes line in the FCG is very stable, and there are hardly any intensity fluctuations in the time domain. However, the intensity fluctuations of the Stokes light appear and are gradually exacerbated as the order increases, which is consistent with the experimental results of the noise measurements. At the same time, the initial symmetric Gaussian distribution of the Stokes light temporal intensity trace becomes gradually asymmetric and eventually evolves to an approximate gamma distribution as the order increases. In fact, the main reason for the deterioration of the laser quality of higher-order Stokes light is none other than the transfer and accumulation of noise in the SBS process among different orders of Stokes and anti-Stokes lines and the extra noise introduction in the sub-cavity. In addition, with the introduction and accumulation of noise, the stability as well as the coherence of the higher-order Stokes light are reduced, making the Stokes light generated by the SBS process at this order more difficult to be resonated in the main cavity, which further deteriorates the coherence of the Stokes light at the next higher order. Thus, the time-domain intensity distribution of these Stokes lines has low stability, and coherence is similar to that of the Stokes light produced in the non-resonant SBS process. Moreover, a similar phenomenon is observed for the anti-Stokes lights of order one to four as illustrated in Figure 14, and the overdependence of the anti-Stokes lines on the parametric gain of the FWM leads to a faster stability deterioration of its temporal trace with order increase than that of the Stokes lines.

In summary, the data analysis shows that the laser quality of each order of Stokes and anti-Stokes light is different in the frequency combs generated by the proposed FCG based on SBS and FWM processes, and the laser quality will deteriorate with the order of SBS process, which is an issue worth being noted and resolved in future work.

4. Discussion

For laser-based frequency comb generators, the number, OSNR, power derivation of tooth and the stability and coherence of each tooth are critical. MW-BRFL with narrow bandwidth, inhomogeneity, high gain coefficient and efficient linewidth suppression characteristics of Brillouin gain and noise suppression capability of random distributed feedback provided by Rayleigh scattering can achieve narrow linewidth, low noise and relatively stable multi-wavelength laser output. In the recently published work, several typical MW-BRFLs have been implemented and characterized. In reference [59], the first MW-BRFL with a dual-cavity structure was proposed, where a 25 km SMF and a 5 km non-uniform fiber were used to provide Brillouin gain and random distributed feedback, respectively. Due to the low gain coefficient of the SMF and the limitation of the EDFA output power, multi-wavelength laser output with only six orders of Stokes lines was finally achieved. Subsequently, to enhance the stability of multi-wavelength laser output, a dual-cavity MW-BRFL based on a polarization-maintaining fiber was proposed to suppress noise and stabilize laser output by eliminating random changes in the polarization state of light in the resonant cavity [60,61]. In addition, another MW-BRFL based on random fiber grating is proposed. With the strong Rayleigh scattering effect in the centimeter-long random fiber grating, the number of generated Stokes lines in the laser output is raised to 14, while ensuring low noise and stability of the laser output [62]. However, the multi-wavelength laser output achieved by these efforts is bound to be bottlenecked by gain and pump power limitations, regardless of the method used to boost the number of Stokes lines. In this work, the introduction of the FWM effect in the fiber allows the anti-Stokes light, which has always been buried in the bottom noise, to be enabled and amplified, which undoubtedly increases the number of lasing lines in the laser output exponentially, without the need for excessive pump power. This allows this experiment to achieve a stable laser output with 17 orders of Stokes lines and 15 orders of anti-Stokes lines. Although the FWM effect in optical fibers is common, the present work provides an unprecedented and significant study on the properties of resonant anti-Stokes lines in MW-BRFL based on this effect.

It is worth noting that this experiment has proposed a novel MW-BRFL while there are still some points to be optimized. These include the enhancement of the Stokes line flatness, the laser output power, the number of Stokes/anti-Stokes lines, and the stability and linewidth of each order of Stokes/anti-Stokes lines. The improvement for line flatness can be achieved by optimizing the power-averaging capability of the NOLM by better balancing the relationship between its rate of transmissivity variation with incident optical power and the additional losses it introduces. The output power of the laser and the number of generated resonant lines can be achieved by the use of high-gain EDFA and higher-nonlinearity HNLF, in addition to optimizing the system structure and reducing intracavity losses. The optimization of the linewidth and stability of each order of Stokes/anti-Stokes lines needs to filter out the redundant random resonant modes within the Brillouin gain bandwidth by using a narrow-bandwidth optical fiber filter to compress the mode competition and multi-mode resonance.

5. Conclusions

In conclusion, a frequency comb generator was experimentally demonstrated based on the dual-cavity Brillouin random oscillator including a main cavity and a sub-cavity. The half-open main cavity incorporates a section of HNLF with a high nonlinearity coefficient ($\geq$10 $W^{-1}\cdot km^{-1}$ @1550 nm) as Brillouin gain fiber, Rayleigh scattering fiber to provide random distributed feedback and the NOLM as a power balancer to obtain the Brillouin random resonance. The sub-cavity includes an EDFA to boost and couple back the low-order Stokes light for the cascade SBS process. Furthermore, the FWM effect in the HNLF is effectively excited under phase-matching conditions, which not only increases the number of frequency combs but also boosts the power of the multi-order Stokes and anti-Stokes light. The NOLM with its unique transmission spectrum is used to balance the power between the different orders of Stokes and anti-Stokes light to achieve a smaller power difference. Eventually, 17 orders of stable Stokes lines and 15 orders of stable anti-Stokes lines within a 10 dB power deviation were obtained with the minimum OSNR of ~10 dB and ~7.5 dB, respectively. In addition, the characteristics of FCG have been experimentally investigated. Both Stokes and anti-Stokes light exhibit the linewidths on the order of kHz or even sub-kHz and an inter-peak spacing of 0.076 nm. As the order of Stokes and anti-Stokes light increases, both RIN and frequency noise gradually accumulate, showing a tendency to become progressively more severe. Importantly, the properties of the anti-Stokes line in the generated frequency comb were studied for the first time. Experiments have demonstrated that the regulation of the power ratio of the pump light to the sub-cavity feedback light during the generation of the frequency combs deserves attention. Such an FCG with fixed frequency spacing will find promising applications in fields of optical communications, microwaves, optical sensing, etc.

Author Contributions: Conceptualization, Y.P. and Y.X.; methodology, Y.P. and Y.X.; software, Y.P.; validation, Y.P.; formal analysis, Y.P.; investigation, Y.P., S.M., Q.J. and Y.X.; resources, Y.X.; data curation, Y.P., S.M. and Q.J.; writing—original draft preparation, Y.P.; writing—review and editing, Y.P. and Y.X.; visualization, Y.P.; supervision, Y.X., X.Z., Y.L., Z.Q. and Z.L.; project administration, Y.X.; funding acquisition, Y.X. All authors have read and agreed to the published version of the manuscript.

Funding: This research was funded by National Natural Science Foundation of China (62105180), Natural Science Foundation of Shandong Province (ZR2020MF110, ZR2020MF118), Taishan Scholar Foundation of Shandong Province (tsqn202211027) and Qilu Young Scholar Program of Shandong University.

Institutional Review Board Statement: Not applicable.

Informed Consent Statement: Not applicable.

Data Availability Statement: Data underlying the results presented in this paper are not publicly available at this time but may be obtained from the authors upon reasonable request.

Conflicts of Interest: The authors declare no conflict of interest.

References

1. Veselka, J.J.; Korotky, S.K. A Multiwavelength Source Having Precise Channel Spacing for WDM Systems. *IEEE Photonics Technol. Lett.* **1998**, *10*, 958–960. [CrossRef]
2. Peng, P.-C.; Tseng, H.-Y.; Chi, S. Long-Distance FBG Sensor System Using a Linear-Cavity Fiber Raman Laser Scheme. *IEEE Photonics Technol. Lett.* **2004**, *16*, 575–577. [CrossRef]
3. Yao, J. Microwave Photonics. *J. Light. Technol.* **2009**, *27*, 314–335. [CrossRef]
4. Shahi, S.; Harun, S.W.; Ahmad, H. Multi-Wavelength Brillouin Fiber Laser Using Brillouin-Rayleigh Scatterings in Distributed Raman Amplifier. *Laser Phys. Lett.* **2009**, *6*, 737. [CrossRef]
5. Wang, F.; Gong, Y. Tunable and Switchable Multi-Wavelength Erbium-Brillouin Random Fiber Laser Incorporating a Highly Nonlinear Fiber. *J. Light. Technol.* **2020**, *38*, 4093–4099. [CrossRef]
6. Bellemare, A.; Karasek, M.; Rochette, M.; LRochelle, S.; Tetu, M. Room Temperature Multifrequency Erbium-Doped Fiber Lasers Anchored on the ITU Frequency Grid. *J. Light. Technol.* **2000**, *18*, 825–831. [CrossRef]
7. Kim, S.K.; Chu, M.J.; Lee, J.H. Wideband Multiwavelength Erbium-Doped Fiber Ring Laser with Frequency Shifted Feedback. *Opt. Commun.* **2001**, *190*, 291–302. [CrossRef]
8. Zhou, K.; Zhou, D.; Dong, F.; Ngo, N.Q. Room-Temperature Multiwavelength Erbium-Doped Fiber Ring Laser Employing Sinusoidal Phase-Modulation Feedback. *Opt. Lett.* **2003**, *28*, 893–895. [CrossRef] [PubMed]
9. Vasseur, J.; Hanna, M.; Dudley, J.M.; Barry, J.R. Numerical and Theoretical Analysis of an Alternate Multiwavelength Mode-Locked Fiber Laser. *IEEE Photonics Technol. Lett.* **2005**, *17*, 2295–2297. [CrossRef]
10. He, W.; Zhu, L.; Dong, M.; Lou, X.; Luo, F. Wavelength-Switchable and Stable-Ring-Cavity, Erbium-Doped Fiber Laser Based on Mach–Zehnder Interferometer and Tunable Filter. *Laser Phys.* **2018**, *28*, 045104. [CrossRef]
11. Luo, A.-P.; Luo, Z.-C.; Xu, W.-C. Tunable and Switchable Multiwavelength Erbium-Doped Fiber Ring Laser Based on a Modified Dual-Pass Mach-Zehnder Interferometer. *Opt. Lett.* **2009**, *34*, 2135–2137. [CrossRef]
12. Ma, L.; Kang, Z.; Qi, Y.; Jian, S. Tunable Dual-Wavelength Fiber Laser Based on an MMI Filter in a Cascaded Sagnac Loop Interferometer. *Laser Phys.* **2014**, *24*, 045102. [CrossRef]
13. Xu, Z.W.; Zhang, Z.X. All-Normal-Dispersion Multi-Wavelength Dissipative Soliton Yb-Doped Fiber Laser. *Laser Phys. Lett.* **2013**, *10*, 085105. [CrossRef]

14. Jin, X.; Lou, Z.; Zhang, H.; Xu, J.; Zhou, P.; Liu, Z. Random Distributed Feedback Fiber Laser at 2.1 μm. *Opt. Lett.* **2016**, *41*, 4923–4926. [CrossRef]
15. Wang, L.; Dong, X.; Shum, P.P.; Su, H. Tunable Erbium-Doped Fiber Laser Based on Random Distributed Feedback. *IEEE Photonics J.* **2014**, *6*, 1–5. [CrossRef]
16. Ma, R.; Quan, X.; Wu, H.; Gao, W.; Huang, D.; Wang, X.; Xu, S.; Fan, D.; Liu, J. 20 Watt-Level Single Transverse Mode Narrow Linewidth and Tunable Random Fiber Laser at 1.5 Mm Band. *Opt. Express* **2022**, *30*, 28795–28804. [CrossRef] [PubMed]
17. Xu, Y.; Zhang, L.; Lu, P.; Mihailov, S.; Chen, L.; Bao, X. Time-Delay Signature Concealed Broadband Gain-Coupled Chaotic Laser with Fiber Random Grating Induced Distributed Feedback. *Opt. Laser Technol.* **2019**, *109*, 654–658. [CrossRef]
18. Pang, M.; Bao, X.; Chen, L. Observation of Narrow Linewidth Spikes in the Coherent Brillouin Random Fiber Laser. *Opt. Lett.* **2013**, *38*, 1866–1868. [CrossRef]
19. Zhang, L.; Xu, Y.; Gao, S.; Saxena, B.; Chen, L.; Bao, X. Linearly Polarized Low-Noise Brillouin Random Fiber Laser. *Opt. Lett.* **2017**, *42*, 739–742. [CrossRef]
20. Wu, H.; Han, B.; Liu, Y. Tunable Narrowband Cascaded Random Raman Fiber Laser. *Opt. Express* **2021**, *29*, 21539–21550. [CrossRef]
21. Xu, Y.; Lu, P.; Bao, X. Compact Single-End Pumped Brillouin Random Fiber Laser with Enhanced Distributed Feedback. *Opt. Lett.* **2020**, *45*, 4236–4239. [CrossRef]
22. Xu, Y.; Zhang, L.; Chen, L.; Bao, X. Single-Mode SOA-Based 1kHz-Linewidth Dual-Wavelength Random Fiber Laser. *Opt. Express* **2017**, *25*, 15828–15837. [CrossRef] [PubMed]
23. Tovar, P.; Temporão, G.; Weid, J.P. von der Longitudinal Mode Dynamics in SOA-Based Random Feedback Fiber Lasers. *Opt. Express* **2019**, *27*, 31001–31012. [CrossRef] [PubMed]
24. Shawki, H.; Kotb, H.; Khalil, D. Modeling and Characterization of a Dual-Wavelength SOA-Based Single Longitudinal Mode Random Fiber Laser with Tunable Separation. *OSA Contin.* **2019**, *2*, 358–369. [CrossRef]
25. Fan, M.; Wang, Z.; Wu, H.; Sun, W.; Zhang, L. Low-Threshold, High-Efficiency Random Fiber Laser with Linear Output. *IEEE Photonics Technol. Lett.* **2015**, *27*, 319–322. [CrossRef]
26. Deng, J.; Han, M.; Xu, Z.; Du, Y.; Shu, X. Stable and Low-Threshold Random Fiber Laser via Anderson Localization. *Opt. Express* **2019**, *27*, 12987–12997. [CrossRef]
27. Wu, H.; Wang, Z.; Sun, W.; He, Q.; Wei, Z.; Rao, Y.-J. 1.5 μm Low Threshold, High Efficiency Random Fiber Laser with Hybrid Erbium–Raman Gain. *J. Light. Technol.* **2018**, *36*, 844–849. [CrossRef]
28. Wang, Z.; Wu, H.; Fan, M.; Zhang, L.; Rao, Y.; Zhang, W.; Jia, X. High Power Random Fiber Laser with Short Cavity Length: Theoretical and Experimental Investigations. *IEEE J. Sel. Top. Quantum Electron.* **2015**, *21*, 10–15. [CrossRef]
29. Zhang, H.; Huang, L.; Song, J.; Wu, H.; Zhou, P.; Wang, X.; Wu, J.; Xu, J.; Wang, Z.; Xu, X.; et al. Quasi-Kilowatt Random Fiber Laser. *Opt. Lett.* **2019**, *44*, 2613–2616. [CrossRef]
30. Ma, R.; Li, J.Q.; Guo, J.Y.; Wu, H.; Zhang, H.H.; Hu, B.; Rao, Y.J.; Zhang, W.L. High-Power Low Spatial Coherence Random Fiber Laser. *Opt. Express* **2019**, *27*, 8738–8744. [CrossRef]
31. Babin, S.A.; El-Taher, A.E.; Harper, P.; Podivilov, E.V.; Turitsyn, S.K. Tunable Random Fiber Laser. *Phys. Rev. A* **2011**, *84*, 021805. [CrossRef]
32. Zhang, Y.; Song, J.; Ye, J.; Xu, J.; Yao, T.; Zhou, P. Tunable Random Raman Fiber Laser at 1.7 μm Region with High Spectral Purity. *Opt. Express* **2019**, *27*, 28800–28807. [CrossRef] [PubMed]
33. Du, X.; Zhang, H.; Wang, X.; Zhou, P. Tunable Random Distributed Feedback Fiber Laser Operating at 1 μm. *Appl. Opt.* **2015**, *54*, 908–911. [CrossRef] [PubMed]
34. Ye, J.; Xu, J.; Zhang, H.; Zhou, P. Powerful Narrow Linewidth Random Fiber Laser. *Photonic Sens.* **2017**, *7*, 82–87. [CrossRef]
35. Pang, Y.; Ma, S.; Zhao, X.; Qin, Z.; Liu, Z.; Xu, Y. Single-Longitudinal-Mode Short-Cavity Brillouin Random Fiber Laser via Frequency Auto-Tracking with Unpumped-EDF Sagnac Loop. *Infrared Phys. Technol.* **2022**, *127*, 104461. [CrossRef]
36. Xu, Y.; Gao, S.; Lu, P.; Mihailov, S.; Chen, L.; Bao, X. Low-Noise Brillouin Random Fiber Laser with a Random Grating-Based Resonator. *Opt. Lett.* **2016**, *41*, 3197–3200. [CrossRef] [PubMed]
37. Pang, Y.; Xu, Y.; Zhao, X.; Qin, Z.; Liu, Z. Low-Noise Brillouin Random Fiber Laser with Auto-Tracking Dynamic Fiber Grating Based on a Saturable Absorption Ring. *Infrared Phys. Technol.* **2022**, *122*, 104088. [CrossRef]
38. Pang, Y.; Xu, Y.; Zhao, X.; Qin, Z.; Liu, Z. Low-Noise Narrow-Linewidth Brillouin Random Fiber Laser with Dynamic Fiber Grating. In Proceedings of the Conference on Lasers and Electro-Optics (2022), Paper JW3B.34, San Jose, CA, USA, 15 May 2022; Optica Publishing Group: Washington, DC, USA, 2022; p. JW3B.34.
39. Du, X.; Zhang, H.; Xiao, H.; Zhou, P.; Liu, Z. Temporally Stable Random Fiber Laser Operates at 1070 Nm. *IEEE Photonics J.* **2015**, *7*, 1–7. [CrossRef]
40. Pang, Y.; Xu, Y.; Zhao, X.; Qin, Z.; Liu, Z. Stabilized Narrow-Linewidth Brillouin Random Fiber Laser with a Double-Coupler Fiber Ring Resonator. *J. Light. Technol.* **2022**, *40*, 2988–2995. [CrossRef]
41. Xu, Y.; Xiang, D.; Ou, Z.; Lu, P.; Bao, X. Random Fabry–Perot Resonator-Based Sub-KHz Brillouin Fiber Laser to Improve Spectral Resolution in Linewidth Measurement. *Opt. Lett.* **2015**, *40*, 1920. [CrossRef]
42. Gao, S.; Zhang, L.; Xu, Y.; Lu, P.; Chen, L.; Bao, X. Tapered Fiber Based Brillouin Random Fiber Laser and Its Application for Linewidth Measurement. *Opt. Express* **2016**, *24*, 28353. [CrossRef] [PubMed]

43. Feng, X.; Lu, C.; Tam, H.Y.; Wai, P.K.A. Reconfigurable Microwave Photonic Filter Using Multiwavelength Erbium-Doped Fiber Laser. *IEEE Photon. Technol. Lett.* **2007**, *19*, 1334–1336. [CrossRef]
44. Shen, Y.; Wang, R.; Pu, T. A Novel Technique to Generate High-Frequency Microwave Signal Based on High-Order Stimulated Brillouin Scattering. *Acta Opt. Sin.* **2010**, *30*, 1571–1575. [CrossRef]
45. Oehler, A.E.H.; Zeller, S.C.; Weingarten, K.J.; Keller, U. Broad Multiwavelength Source with 50 GHz Channel Spacing for Wavelength Division Multiplexing Applications in the Telecom C Band. *Opt. Lett.* **2008**, *33*, 2158. [CrossRef] [PubMed]
46. Oh, S.H.; Shin, J.-U.; Park, Y.-J.; Kim, S.-B.; Park, S.; Sung, H.-K.; Baek, Y.-S.; Oh, K.-R. Multiwavelength Lasers for WDM-PON Optical Line Terminal Source by Silica Planar Lightwave Circuit Hybrid Integration. *IEEE Photon. Technol. Lett.* **2007**, *19*, 1622–1624. [CrossRef]
47. Xu, Y.; Lu, P.; Mihailov, S.; Bao, X. Real-Time Physical Random Bit Generation at Gbps Based on Random Fiber Lasers. *Opt. Lett.* **2017**, *42*, 4796–4799. [CrossRef]
48. Zhu, T.; Ke, T.; Rao, Y.; Chiang, K.S. Fabry–Perot Optical Fiber Tip Sensor for High Temperature Measurement. *Opt. Commun.* **2010**, *283*, 3683–3685. [CrossRef]
49. Han, Y.-G.; Tran, T.V.A.; Kim, S.-H.; Lee, S.B. Multiwavelength Raman-Fiber-Laser-Based Long-Distance Remote Sensor for Simultaneous Measurement of Strain and Temperature. *Opt. Lett.* **2005**, *30*, 1282. [CrossRef]
50. Xu, Y.; Zhang, M.; Lu, P.; Mihailov, S.; Bao, X. Multi-Parameter Sensor Based on Random Fiber Lasers. *AIP Adv.* **2016**, *6*, 095009. [CrossRef]
51. Xu, Y.; Zhang, M.; Lu, P.; Mihailov, S.; Bao, X. Multi-Parameter Fiber-Optic Sensors Based on Fiber Random Grating. In Proceedings of the 2017 25th Optical Fiber Sensors Conference (OFS), Jeju, Republic of Korea, 24–28 April 2017; pp. 1–4.
52. Xu, Y.; Zhang, L.; Gao, S.; Lu, P.; Mihailov, S.; Bao, X. Highly Sensitive Fiber Random-Grating-Based Random Laser Sensor for Ultrasound Detection. *Opt. Lett.* **2017**, *42*, 1353–1356. [CrossRef] [PubMed]
53. Xu, Y.; Lu, P.; Gao, S.; Xiang, D.; Lu, P.; Mihailov, S.; Bao, X. Optical Fiber Random Grating-Based Multiparameter Sensor. *Opt. Lett.* **2015**, *40*, 5514–5517. [CrossRef]
54. Wu, H.; Han, B.; Wang, Z.; Genty, G.; Feng, G.; Liang, H. Temporal Ghost Imaging with Random Fiber Lasers. *Opt. Express* **2020**, *28*, 9957–9964. [CrossRef] [PubMed]
55. Liu, Y.W.; Wang, Z.; Wang, S.S.; Ma, R.; Zhang, Y.; Zhang, W.L. Novel Illumination for Imaging Using Self-Modulated Coherent Random Fiber Laser. In Proceedings of the Conference on Lasers and Electro-Optics (2022), Paper JW3B.192, San Jose, CA, USA, 15 May 2022; Optica Publishing Group: Washington, DC, USA, 2022; p. JW3B.192.
56. Gao, S.; Zhang, L.; Xu, Y.; Chen, L.; Bao, X. High-Speed Random Bit Generation via Brillouin Random Fiber Laser with Non-Uniform Fibers. *IEEE Photon. Technol. Lett.* **2017**, *29*, 1352–1355. [CrossRef]
57. Xiang, D.; Lu, P.; Xu, Y.; Gao, S.; Chen, L.; Bao, X. Truly Random Bit Generation Based on a Novel Random Brillouin Fiber Laser. *Opt. Lett.* **2015**, *40*, 5415. [CrossRef]
58. Harun, S.W.; Shirazi, M.R.; Ahmad, H. A New Configuration of Multi-Wavelength Brillouin Fiber Laser. *Laser Phys. Lett.* **2008**, *5*, 48–50. [CrossRef]
59. Zhang, L.; Xu, Y.; Gao, S.; Saxena, B.; Chen, L.; Bao, X. Multiwavelength Coherent Brillouin Random Fiber Laser with Ultrahigh Optical Signal-to-Noise Ratio. *IEEE J. Sel. Top. Quantum Electron.* **2018**, *24*, 1–8. [CrossRef]
60. Zhang, L.; Wang, Y.; Xu, Y.; Chen, L.; Bao, X. Linearly Polarized Multi-Wavelength Comb via Rayleigh Scattering Induced Brillouin Random Lasing Resonance. In Proceedings of the 2018 Optical Fiber Communications Conference and Exposition (OFC), San Diego, CA, USA, 11–15 March 2018; pp. 1–3.
61. Zhang, L.; Wang, Y.; Xu, Y.; Zhou, D.; Chen, L.; Bao, X. Linearly Polarized Multi-Wavelength Fiber Laser Comb via Brillouin Random Lasing Oscillation. *IEEE Photonics Technol. Lett.* **2018**, *30*, 1005–1008. [CrossRef]
62. Zhang, L.; Xu, Y.; Lu, P.; Mihailov, S.; Chen, L.; Bao, X. Multi-Wavelength Brillouin Random Fiber Laser via Distributed Feedback From a Random Fiber Grating. *J. Light. Technol.* **2018**, *36*, 2122–2128. [CrossRef]
63. Wang, F.; Gong, Y. Switchable Wide-Band Multi-Wavelength Brillouin-Erbium Fiber Laser Based on Random Distributed Feedback. *IEEE Access* **2020**, *8*, 218614–218620. [CrossRef]

hv ***photonics*** MDPI

Article

Chiral Dual-Core Photonic Crystal Fiber for an Efficient Circular Polarization Beam Splitter

She Li [1,2,*], Yibing Li [1], Hongwei Lv [3], Changtong Ji [2], Hongze Gao [4] and Qian Sun [1]

1 College of Information and Communication Engineering, Harbin Engineering University, Harbin 150001, China
2 College of Science, Heilongjiang University of Science and Technology, Harbin 150022, China
3 The Fourth Affiliated Hospital, Harbin Medical University, Harbin 150001, China
4 School of Instrument Science and Engineering, Harbin Institute of Technology, Harbin 150001, China
* Correspondence: lishe1979@163.com

Abstract: As a function of a circular polarization beam splitter (CPBS), combining a linear polarization beam splitter with a quarter-wave plate results in a polarization error in a circular polarization fiber-optic circuit. To relieve the error, chiral dual-core photonic crystal fiber (DC-PCF) is investigated as a kind of an efficient circular polarization beam splitter by using the chiral plane-wave expansion (PWE) method. On the basis of the competitive effect in polarization and coupling length between the circular asymmetry of the structure and the chirality of the medium, the effects of the structure and the chirality are analyzed. The numerical results demonstrate that a CPBS needs the weak circular asymmetry in its structure and a relatively stronger chirality of the medium. Then, a kind of CPBS based on chiral DC-PCF is designed with weaker chirality, with a central wavelength of 1.55 μm. The simulation shows the superior performance of having a shorter coupling length and a higher extinction ratio. Furthermore, the dual-wavelength of 1.55 μm and 1.30 μm with left-circular polarization can further be separated by the corresponding chiral DC-PCF. The results show promising applications for the circular polarized multiplexer/demultiplexer in fiber laser communication systems.

Keywords: chiral photonic crystal fiber; circular polarization beam splitter; fiber-optical element

Citation: Li, S.; Li, Y.; Lv, H.; Ji, C.; Gao, H.; Sun, Q. Chiral Dual-Core Photonic Crystal Fiber for an Efficient Circular Polarization Beam Splitter. *Photonics* **2023**, *10*, 45. https://doi.org/10.3390/photonics10010045

Received: 30 November 2022
Revised: 22 December 2022
Accepted: 28 December 2022
Published: 1 January 2023

1. Introduction

Dual-core photonic crystal fiber (DC-PCF), as a kind of beam splitter that can split beams with different polarizations or wavelengths, has gradually attracted more and more attention thanks to the composition of its having only one material and its unusual properties, such as tailored coupling length (CL), spectral bandwidth and extinction ratio (ER) [1–5], and it can be applied in circuits for fiber laser communication, optical integration and sensing. According to the polarization states of light, the devices can be classified into linearly polarized beam splitters (LPBSs) and circularly polarized beam splitters (CPBSs), which can respectively separate the orthogonally linearly [1–5] and oppositely circularly polarized light [6]. Generally, the LPBSs can be realized by introducing orthogonal asymmetry, named circular asymmetry (geometric anisotropy or distribution of materials), into a cross section of the fiber, such as the air hole's or core's position [1–3], size [2–8], shape [9,10] or holes filled with some liquid or multiple materials [4,11–16], and applied in the linearly polarized system to suppress the polarization error or noise [17,18]. Because of the inherent defects, such as the polarization axis's imperfectly aligning during welding and stress caused by extrusion or stretching in practical applications, the polarization error and noise would inevitably occur [17,18]. Research results demonstrate that polarization error and noise can be suppressed, even eliminated, by using a circularly polarized optical fiber system [19–21], since the circular birefringence can suppress the linear one in spun optical fiber for higher spun rates [22]. However, the polarization axis's imperfectly aligning still remains owing to the use of all-fiber quarter-wave plates and LPBSs as all-fiber CPBSs [19–21]. In fact, in the process of producing circularly polarized light, the quarter-wave plates can also introduce the error because of the imperfect interception; however, the CPBS can split the linearly polarized (LP) light to right-handed circularly polarized (RCP) and left-handed circularly polarized (LCP) ones without any additional devices or procedures because the LP light can be viewed as a combination of RCP and LCP ones. Thus, it is necessary to study all-fiber CPBSs formed by the circular polarization–maintaining fibers.

In general, the circular polarization–maintaining fibers can employ chiral structures [6,22–26] or chiral mediums [27–31] that have axial rotation (twisted or spiral) and/or optical activity [22,23,30,31], namely structurally and dielectrically chiral fiber, where the operation wavelength is far less and much larger than the chiral object in size. The former, which has the helically twisted or spun structure, forms a kind of chiral fiber or chiral fiber grating because of the different twist angle [22–26], which can support one or two-paired circularly polarized modes. Theoretical research results demonstrate that the twisted DC-PCF can be viewed as an all-fiber CPBS, whereas fabricated technology for undergoing symmetric distortions in twisting and drawing poses a challenge [6]. For the latter, theoretical studies have demonstrated that they can support the circularly polarized modes with single- and multicore [27–31], in which a left-handed polarization beam can be leaked by single-polarization single-mode fiber that can be the manufactured as an ordinary PCF [28]. Recently, as a kind of CPBS, the dielectrically chiral DC-PCF can separate the oppositely circularly polarized light with stronger chirality strength [32]; however, its coupling length is very long. Therefore, in this paper, we will focus on the dielectrically chiral DC-PCF CPBS to optimize its structural parameters for short CL and decrease the dependence on the chirality of the medium.

In this paper, as a kind of CPBS, the dielectrically chiral DC-PCF is first considered with a shorter distance between the two cores. Given the application in practice, the operation wavelengths are selected as 1.55 μm and 1.30 μm, which are easy to be emitted by lasers, such as a quantum dot laser, a fiber laser, a superluminescent diode laser, and so on [33–38], and the fiber can achieve propagation with zero dispersion or near-zero dispersion for the two band [39–41]. Using the chiral plane-wave expansion (PWE) method [27], the effects of the circular asymmetry of the structure and the chirality of the medium are analyzed. From the competition between the circular asymmetry of structure and the chirality of the medium in polarization and CL, a shorter PCF CPBS with a weaker chirality is obtained by optimizing the geometry parameters. Then, as a kind of multiplexer/demultiplexer for circularly polarized (CP) beams is discussed, and the numerical results show that it can split beams at the dual-wavelengths of 1.55 μm and 1.30 μm that are widely used in fiber laser communication systems.

2. Model and Analysis

In order to obtain the shorter CL, having strong coupling between cores is necessary [1]. Hence, the chiral DC-PCF with a shorter distance between the two cores is considered, as shown in Figure 1, where the two defects (A and B) denote the dual cores; their distance is 2Λ, and Λ and d, respectively, represent the lattice constant and diameter in the air holes. The darker part is annotated as the chiral medium, and the white holes are the air ones. The tailored holes of the diameter of are, respectively, $D1$ and $D2$, which can determine the modal linearly birefringence of the fiber to tailor the CL and the corresponding polarization states.

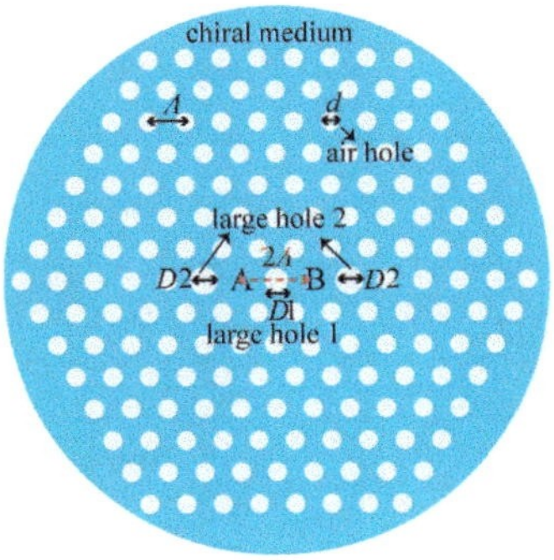

Figure 1. The schematic diagram of a chiral DC-PCF.

Generally, the fiber has high birefringence when $D1/d$ and $D2/d$ are larger/less than unity; the higher the birefringence, the shorter the CLs. Additionally, when $D1/d = 1$ and $D2/d = 1$, the fiber is identical to a conventional DC-PCF, which has slight birefringence because of the asymmetry of the structure. To reduce the dependence on the chirality of the medium, the birefringence should not be too large, because of the competitive effect between the asymmetry of the structure and the chirality of the medium in polarization and CL [32].

The Drude–Born–Fedorov's constitutive relations for the chiral medium, which are employed with $\mathbf{D} = \varepsilon_0\varepsilon_r(\mathbf{E} + \xi\nabla \times \mathbf{E})$ and $\mathbf{B} = \mu_0\mu_r(\mathbf{H} + \xi\nabla \times \mathbf{H})$, are chosen [28]. ξ represents the chirality strength characterized by its specific rotatory power δ of the medium through $\delta = -k^2n^2$, where k and n, respectively, denote the wavenumber in a vacuum and the average refractive index of the medium. Hence, the air can be viewed as a chiral medium with $\delta = 0$ (achiral medium).

The coupling characteristics of a chiral DC-PCF are investigated through the chiral plane-wave expansion (PWE) method [27]

$$\beta^2\sum_{n,\alpha} A_{mn}^{\alpha\gamma}H_n^{\alpha} + \beta\sum_{n,\alpha} B_{mn}^{\alpha\gamma}H_n^{\alpha} + \sum_{n,\alpha} C_{mn}^{\alpha\gamma}H_n^{\alpha} = 0 \tag{1}$$

Because polymethyl methacrylate (PMMA) can be doped with a chiral medium or synthesized with a chiral copolymer polymer used to draw polymer PCF [42–44], PMMA-doped griseofulvin is chosen as the chiral background of the DC-PCF in the simulation [44], where the chirality can be introduced by griseofulvin with the solution-doping technique [43]. Further, more dilute dopant hardly affects the material dispersion of PMMA, so the formula of material dispersion can be described by the pure PMMA:

$$n^2 - 1 = \sum_{i=1}^{3} A_i\lambda^2/(\lambda^2 - l_i^2) \tag{2}$$

where $A_1 = 0.4963$, $l_1 = 71.8$ nm, $A_2 = 0.6965$, $l_2 = 117.4$ nm, $A_3 = 0.3223$ and $l_3 = 9237$ nm [42]. The optical rotatory dispersion could be expressed by the empirical Boltzmann formula: $\delta_0 = B_1/\lambda^2 + B_2/\lambda^4$, where $B_1 = 1.46 \times 10^{4\circ}\cdot\text{nm}^4/\text{mm}$ and $B_2 = 1.82 \times 10^{10\circ}\cdot\text{nm}^4/\text{mm}$ are employed [44]. In this paper, the interesting wavelength is fixed at 1.55 μm, and the corresponding specific rotatory power is $\delta_0 = 0.0092^\circ/\text{mm}$.

The normalized third Stokes parameter S with weighted intensity is selected for the chiral single-core PCFs [28,29]:

$$S_3 = \iint_{core} s_3|E_t|^2 dS \Big/ \iint_{core} |E_t|^2 dS \tag{3}$$

S_3 ranges from -1 to +1, where negative and positive signs, respectively, indicate right-handed polarized (RHP) and left-handed polarized (LHP) states. The $|S_3|$ closer to unity means purer right-/left-handed CP (RCP/LCP) ones, and S_3 = 0 represents the linearly polarized (LP) one. To ensure that the modes are CP, for convenience, the value of S_3 corresponding to the CL is defined as:

$$S = (S_{3ei} + S_{3oi})/2 \quad i = r, l \tag{4}$$

where r and l, respectively, mean RHP and LHP modes and S_{3ei} and S_{3oi}, respectively, denote the polarization states of the even and odd i-modes. The $|S|$ for the traditionally circular polarization–maintaining fiber approximately equals to 0.96; thus, the case of $|S| > 0.96$ is viewed as a purer CP state, in this paper [26].

The CL with i-modes of chiral DC-FCF is defined as [6,32]

$$L_{Ci} = \frac{\pi}{|\beta_{ei} - \beta_{oi}|} = \frac{\lambda}{2|n_{ei} - n_{oi}|} \tag{5}$$

where β_{ei}, β_{oi} and n_{ei}, n_{oi}, respectively, indicate the paired propagation constants and the corresponding effective indices of the even and odd i-modes. λ is the working wavelength of light in a vacuum.

Because the length of the coupler is short, the transmission loss can be ignored. According to the coupling equation, when a fundamental i-mode power P_{Ai_in} is inputted into core A, after propagating a distance L, the output powers P_{Ai_out} in the output port A; thus, normalized output powers P_{Ai} can be calculated as follows [6,32]:

$$P_{Ai} = \frac{P_{Ai\_out}}{P_{Ai\_in}} = \cos^2\left(\frac{\pi L}{2L_{Ci}}\right) \tag{6}$$

In order to completely split the RHP and LHP light, the total physical length L of the PBS must satisfy the sufficient condition $L = mL_{Cr} = nL_{Cl}$, where L_{Cr} and L_{Cl}, respectively, represent the CLs for the RHP and LHP modes and where m and n are positive integers with opposite parity. Therefore, the coupling length ratio (CLR) is defined as follows [6,32]:

$$CLR = \frac{L_{Cr}}{L_{Cl}} = \frac{n}{m} \tag{7}$$

From the above analysis, it is important to obtain an appropriate CLR. The extinction ratio (ER) characterizes the split ability of light for opposite handedness at the identical output port when both the RCP and LCP beams have the same power incidence to the identical input port [6,32], and it is defined as

$$ER = \left|10\lg\frac{P_{Ar}}{P_{Al}}\right| \tag{8}$$

where P_{Ar} and P_{Ai} are, respectively, the normalized output powers for RHP and LHP light in the identical output port. The larger the value of $|ER|$, the more thorough the split ability for RHP and LHP light and the better performance of the PBS. The $|ER| = 0$ indicates that they cannot be separated at all, because the output powers at the identical output port for the two polarized light are the same.

3. Simulation and Discussion

Generally, the lattice constant determines the spacing between the two cores and plays an essential role in the coupling effect. Thus, the effects of the lattice constant are first investigated, as shown in Figure 2, in which d, $D2/d$ and δ are employed as 0.4Λ, 1 and 1000δ_0, respectively, and $D1/d$ are, respectively, chosen as 1.4, 1.5 and 1.6. In the figure, the same color denotes the characteristics for the fiber with an identical $D1/d$, in which the solid and dotted lines respectively indicate the RHP and LHP modes.

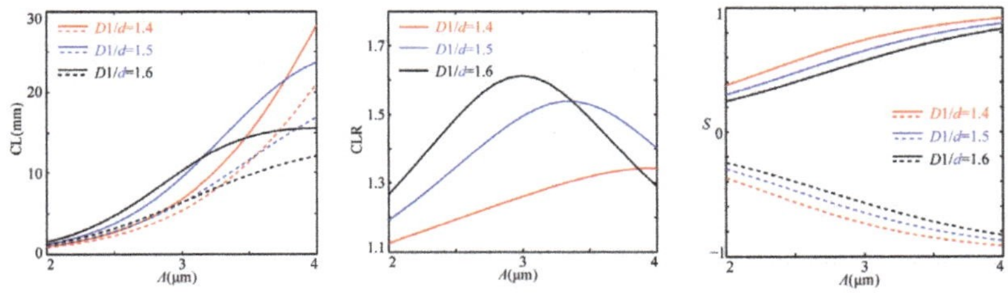

(a) **(b)** **(c)**

Figure 2. The variation relations of CLs (**a**), CLRs (**b**) and Ss (**c**) with Λ.

It is clear that all the CLs lengthen with an increase in the Λ (as reference [32]), in which the CL of *r*-modes is longer than that of the *l*-modes for the paired CLs owing to the right chirality of the medium, and all their differences first increase and then decrease. In the region of the small Λ, the paired CLs lengthen and the difference enlarges with an increase in $D1/d$ or Λ thanks to the effect of the (orthogonal) asymmetry. As Λ increases, the CLs lengthen, which is caused by the larger space between the two cores, and because of the circular asymmetry of the cross section, the difference further enlarges. Additionally, with a further increase in Λ, all the mode fields would be localized in the core area where the effect of the orthogonal asymmetry reduces; the effect of chirality effect relatively strengthens; and the paired CLs and their difference lessens, in which the paired CLs decrease and increase for *r*- and *l*-modes, respectively. As shown in Figure 2b, the corresponding CLRs first increase and then decrease, in which the maximum values of 1.341, 1.538 and 1.612 emerge at Λ = 4.0 μm, Λ = 3.3 μm and Λ = 3.0 μm for $D1/d$ = 1.4, $D1/d$ = 1.5 and $D1/d$ = 1.6, respectively. A comparison of Figure 2b,c shows that the corresponding S can be obtained with 0.9166/−0.9161, 0.7393/−0.7385 and 0.5726/−0.5719, which means that the larger circular asymmetry of the structure introduced by the larger $D1/d$ leads to a reduction in the paired $|S|$ and an increase in their difference, whereas the paired CLs relatively shorten for the larger $D1/d$.

In addition, the $|S|$ rise with an increase in the Λ or a decrease in the $D1/d$. This demonstrates that the realization of CPBS requires increasing the chirality of the medium or decreasing the circular asymmetry of the structure to find a moderate value, such as $D1/d$ = 1.4, where L_{Cr} = 28.30 mm and L_{Cl} = 21.10 mm, in this case, such that the corresponding CLR can approach to 4/3 by further increasing Λ or δ and ensure that $|S| > 0.96$, and the corresponding whole length of the PBS may be 80 mm or longer. Thus, its length would be longer than the ordinary LPBS unless there is a medium with stronger chirality and a larger circular asymmetry in its structure, such as the larger $D1/d$.

Because the closer the core spacing is, the shorter the CL is and the weaker the effect of chirality [32], a smaller lattice constant with Λ = 2.0 μm and stronger chirality with $\delta = 5000\delta_0$ are considered, as shown in Figure 3, in which the relations of CLs (a), CLRs (b) and Ss (c) with $D1/d$ are exhibited, where d is 0.4Λ and $D2/d$ are, respectively, chosen as 1, 1.4 and 1.6. The same color denotes the fiber with identical $D2/d$, in which the solid and dotted lines respectively indicate the *r*- and *l*-modes.

All the paired CLs lengthen with an increase in the $D1/d$ and a decrease in the $D2/d$, and their tendencies are the same, in which the CL of the *r*-modes is longer than that of the corresponding *l*-modes (see Figure 2a); the paired CLs first bifurcate slowly thanks to the smaller circular asymmetry of the structure ($D1/d < 1.2$), and then, with an increase in the circular asymmetry of the structure ($D1/d > 1.2$), the bifurcations rapidly enlarge and decrease after they have reached the maximum. As shown in Figure 3b, the CLRs first slowly rise in the range of $D1/d < 1.2$, then increase rapidly within $1.2 < D1/d < 1.6$ and decrease in the case of $1.75 < D1/d$.

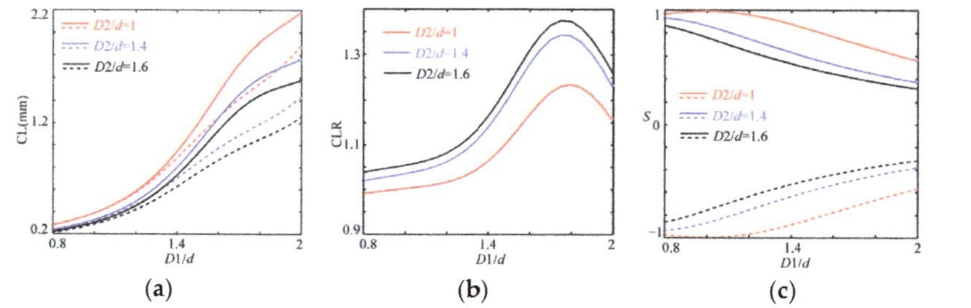

Figure 3. The variation relations of CLs (**a**), CLRs (**b**) and Ss (**c**) with $D1/d$.

From Figure 3c, one can see that all the $|S|$ always lessens with an increase in $D1/d$, except in the case of $D1/d = 1$, in which $|S| = 1$ at $D1/d = 1$ indicates the weakest circular asymmetry of the structure, and whether $D1/d > 1$ or $D1/d < 1$, the value of $D1/d$ away from unity means that the circular asymmetry of the structure strengthens. Thus, all the CLRs are close to unity and the corresponding $|S|$ approaches unity at $D1/d < 1.2$, which indicates that the CPBS can be realized and its physical length would be very long and that the other region needs a stronger chirality for the medium. For example, the maximum values of CLR with 1.236, 1.362 and 1.376 for $D2/d = 1$, $D2/d = 1.4$ and $D2/d = 1.6$ occur at $D1/d \approx 1.80$, $D1/d \approx 1.75$ and $D1/d \approx 1.75$, so by comparing them to Figure 3c, the S approximately equals to $0.6661/-0.6670$, $0.4276/-0.4250$ and $0.3911/-0.3886$, respectively. This demonstrates that the CPBS can be designed by reducing $D1/d$ and $D2/d$ or increasing Λ and δ. Meanwhile, this suggests that the designed CPBS needs the value of CLR on the left of the maximum value because its $|S|$ is closer to unity.

Figure 4 shows the effect of $D2/d$ on CLs (a), CLRs (b) and Ss (c), where Λ, $D1/d$ and δ are chosen as 2 μm, 1.4 and $5000\delta_0$, respectively, and d/Λ are 0.4, 0.42 and 0.44. The same color describes the fiber with an identical d/Λ and the solid and dotted lines respectively indicate the *r*- and *l*-modes.

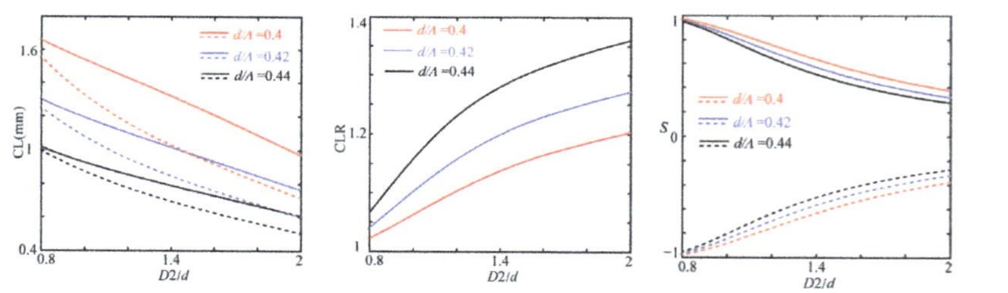

Figure 4. The variation relations of CLs (**a**), CLRs (**b**) and Ss (**c**) with $D2/d$.

All the paired CLs shorten with an increase in $D2/d$ or d/Λ thanks to the circular asymmetry of the structure, and their difference rises, whereas the corresponding CLR increases, as shown in Figure 4a,b. The larger air filling rate would distribute more light into the air holes, which decreases the effective indices and their differences in the paired modes; whereas the light localized in the core area can be concentrated in the center of the core, the circular asymmetry of structure is strengthening thanks to the introduction of the tailoring holes, and with an increase in $D2/d$, the circular asymmetry can further increase; hence, the paired CLs lessen and the CLR rises. This offers a way to reduce the length of the CL and ensure the value of CLR far away from unity to decrease the physical length of the PBS. However, this conflicts with the circular polarization–maintaining property, as shown in Figure 4c, in that with an increase in $D2/d$ or d/Λ, all the $|S|$ are decreasing. The cause is a decrease in the chirality filling rate, as well as the strengthening circular asymmetry of the structure. This demonstrates that designing CPBSs requires increasing the chiral filling rate and decreasing the value of $D2/d$ as it approaches unity, which can ensure that the CLR equals 11/10, 12/11 or some other value close to unity for positive integers with opposite parity to reduce the dependence on the chirality of the medium.

For the case of d/Λ= 0.44 at $D2/d$ = 0.80 and $D2/d$ = 0.85, the CLs for the *r*-/*l*-modes are 1.632 mm/1.497 mm and 1.602 mm/1.438 mm; the corresponding CLRs are 1.090 and 1.114 close to 11/10 and 12/11; and the S are, respectively, 0.9222/−0.9226 and 0.8866/−0.8864. For the case of d/Λ = 0.42 around CLR = 11/10 and CLR = 12/11; the values of CLR are equal to 1.102 and 1.116 at $D2/d$ = 1.00 and $D2/d$ = 1.05, the paired CLs are 1.197 mm/1.087 mm and 1.174 mm/1.051 mm with CLR = 1.102 and CLR = 1.117; and the S are 0.8475/−0.8465 and 0.8116/−0.8099, respectively. Additionally, when d/Λ = 0.40, the paired CLs are 0.8550 mm/0.7751 mm and 0.8382 mm/0.7535 mm for $D2/d$ = 1.15 and $D2/d$ = 1.25, and the S are equal to 0.7912/−0.7892 and 0.7282/−7255.

According to the above analysis, the DC-PCF CPBS may be realized with CLR = 11/10 and CLR = 12/11 by increasing the chirality of the medium or tailoring the structural parameters to reduce the circular asymmetry of the structure. Furthermore, around CLR = 11/10 and CLR = 12/11, to reduce the dependence on the chirality, the CPBS can be easily realized with a larger d/Λ, although its filling rate for chirality is smaller; however, its practical length would be long. For the other value of CLR closer to unity, the required chirality of the medium would be weaker, and the length of the CPBS must be longer. The shorter length of the CPBS can be designed with CLR far away from unity unless the chirality of the medium is strong enough.

Figure 5 shows the effect of the filling rate of air (d/Λ), where a larger Λ = 2.5 μm and $\delta = 5000\delta_0$ are employed to increase the chirality effect, and $D1/d$ and $D2/d$ are, respectively, the same, at 1.4, 1.5 and 1.6.

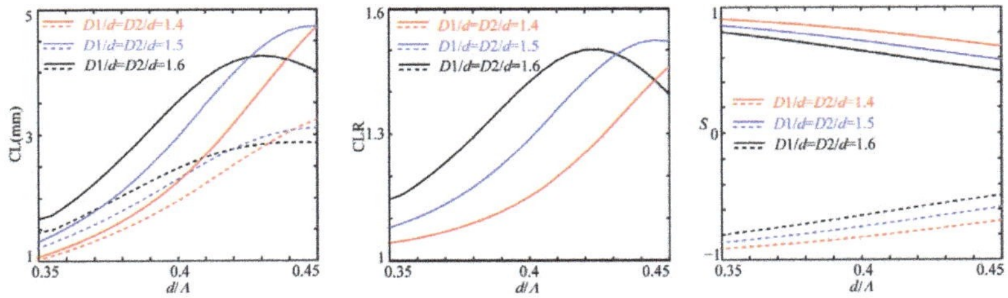

Figure 5. The variation relations of CLs (**a**), CLRs (**b**) and *S*s (**c**) with d/Λ.

In Figure 5a, all the paired CLs have the same variation tendencies, except in the case of $D1/d$ = $D2/d$ = 1.4, thanks to the limitation of the filling rate of air, in which the paired CLs and its differences always rise, and for the other cases, the two-paired CLs and the differences between the paired CLs first increase and then decrease. This is because the larger air holes squeeze the light into the core area to reduce the coupling of the paired modes between the two cores, and the introduction of the tailored holes further enlarges the difference between the paired CLs. By further increasing d/Λ, more light is localized into the core area to further enhance the effect of the tailored holes, such as in Figures 3a and 4a; the differences reduce, as exhibited in Figure 5b; and the lines of the CLRs first rise and then decline.

In addition, with an increase in d/Λ, $D1/d$ or $D2/d$, all the $|S|$ decrease thanks to a reduction in the filling rate of chirality, as shown in Figure 5c. This indicates that the reduction in the circular asymmetry (smaller $D1/d$ or $D2/d$) or the filling rate of air would increase the chirality effect. At d/Λ = 0.36, the CLs are, respectively, 1.215 mm/1.154 mm, 1.519 mm/1.378 mm and 1.885 mm/1.597 mm for the paired *r*-/*l*-modes; for $D1/d$ = $D2/d$ = 1.4, $D1/d$ = $D2/d$ = 1.5 and $D1/d$ = $D2/d$ = 1.6, the CLRs are, respectively, 1.053, 1.103 and 1.181; and the *S*s are, respectively, 0.8968/−0.8950, 0.8405/−0.8382 and 0.7770/−7745. To reduce the dependence on the chirality of the medium, the DC-PCF can be chosen as a CPBS in the case of $D1/d = D2/d = 1.4$, where the CLR can be designed with 21/20 or with the other values of an integer ratio closer to unity with opposite parity, by further increasing the chirality of the medium or optimizing the structural parameters.

Figure 6 shows the effect of all the tailored holes with the same size, in which the larger Λ is chosen as 2.5 μm, 3 μm and 3.5 μm to reduce the dependence on the chirality of the medium, and d and δ are chosen as 0.4Λ and $5000\delta_0$, respectively.

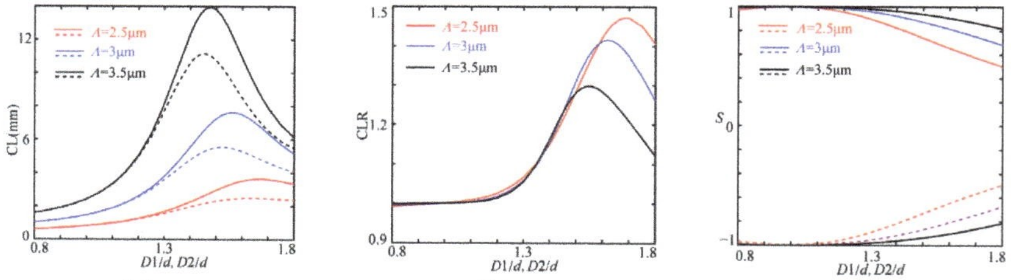

Figure 6. The variation relations of CLs (**a**), CLRs (**b**) and Ss (**c**) with $D1/d$ and $D2/d$.

All the paired CLs and their differences lengthen and then shorten with an increase in $D1/d$ and $D2/d$, and their maximum values reduce and move to the right with a decrease in Λ. Although the maximum differences for the paired CLs decrease, their ratio increases, as shown in Figure 6b, where they are almost equal at $D1/d < 1.5$ and $D2/d < 1.5$, in which they are nearly equal to unity at $D1/d < 1.2$ and $D2/d < 1.2$. This suggests that the design of the DC-PCF for the CPBS have to satisfy the conditions of $D1/d > 1.2$ and $D2/d > 1.2$; otherwise, the CPBS would be long. When the CLRs with maximum values of 1.471, 1.413 and 1.298 respectively occur at $(D1/d=)D2/d = 1.7$, $(D1/d=)D2/d = 1.6$ and $(D1/d=)D2/d = 1.55$ for $\Lambda = 2.5$ μm, $\Lambda = 3$ μm and $\Lambda = 3.5$ μm, the CLs are, respectively, 2.606 mm/2.147 mm, 7.490 mm/5.300 mm and 12.632 mm/9.728 mm and the corresponding Ss are 0.7810/−0.7775, 0.8715/−0.8687 and 0.9382/0.9364, respectively, in which the third case can be viewed as a nearly CP PBS. This indicates that the CPBS can be manufactured by only optimizing the asymmetry of the structure, strengthening the chirality of the medium or increasing the lattice constant, whereas the practical length of the CPBS would be relatively longer.

In Figure 6c, the values of $|S|$ are larger than 0.96, which respectively emerged at $(D1/d=)D2/d = 1.2$, $(D1/d=)D2/d = 1.3$ and $(D1/d=)D2/d = 1.42$, and the corresponding CLRs are, respectively, smaller at 1.026, 1.058 and 1.191, which means that the CLR can be designed for the third case with 6/5 or 7/6 for a stronger chirality effect, and for two other cases, it should have other values closer to unity for an integer ratio with an opposite parity. In conclusion, the larger Λ can increase the chirality effect and reduce the chirality of the medium, but the length of the CPBS would be long.

To show a reduction in the dependence on chirality and how an appropriate value for CLR is obtained, Figure 7 features the CLs, CLR and Ss for the tailored fiber structure, where the larger Λ with 3.5 μm and a smaller d with 0.36 Λ are selected; the larger $D1/d$ with 1.5 and 1.6 are fixed; and δ is $5000\delta_0$.

All the CLs are shortened with a decrease in $D1/d$ and an increase in $D2/d$, which are the same as in Figures 3 and 4. For the case of $D1/d = 1.6$, the CLRs meet the requirement of the design for the CPBS that approach 8/7, 7/6 and 6/5 at $D2/d = 1.412$, $D2/d = 1.540$ and $D2/d = 1.750$. The corresponding CLs are 9.870 mm/8.635 mm, 9.507 mm/8.148 mm and 8.772 mm/7.304 mm, where the Ss are, respectively, 0.9676/−0.9667, 0.9566/−0.9554 and 0.9342/−9326 for the r- and l-modes, which means that a DC-PCF CPBS can be realized with 69.09 mm with CLR = 8/7 at $D2/d = 1.412$. Compared to the previous work, the length of the DC-PCF CPBS is 135 mm shorter and the strength of chirality decreases to 46°/mm [32]. The chirality of the medium can further reduce as long as $|S| \geq 0.96$. For the other two values of CLR, the CPBS can be designed with a shorter length as long as it has stronger chirality.

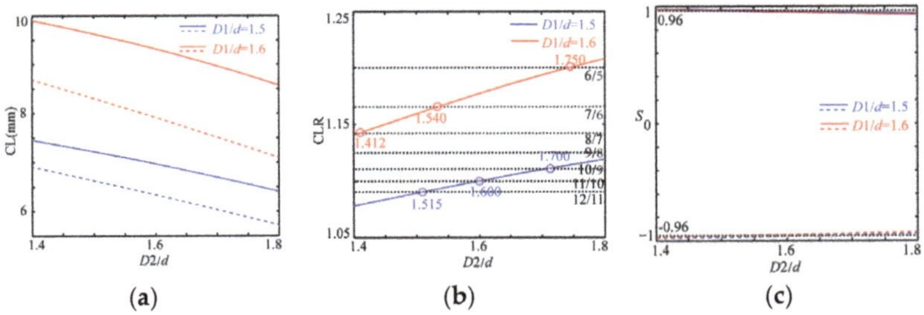

Figure 7. The variation relations of CLs (**a**), CLRs (**b**) and Ss (**c**) with $D2/d$ for fixed $D1/d$.

In the case of $D1/d$ = 1.5, the values of CLRs close to 12/11, 11/10 and 10/9 emerge at $D2/d$ = 1.515, $D2/d$ = 1.600 and $D2/d$ = 1.700, respectively; the lengths of the CLs are, respectively, 7.185 mm/6.585 mm, 6.970 mm/6.335 mm and 6.699 mm/6.034 mm, where the Ss are, respectively, 0.9679/−0.9669, 0.9605/−0.9594 and 0.9511/−0.9498; and thus, the DC-PCF CPBS can be obtained with 79.035 mm for CLR = 12/11 and 69.70 mm for 11/10, which are much less than those in the previous work and need a weaker chirality with 46°/mm [32], where the chirality strength can be furtherly decreased in the case of CLR = 12/11.

Figure 8 shows the normalized transmission power variation for the RHP and LHP modes along the propagation distance in core A and the corresponding ER for the three CPBSs. After a propagation distance of 69.09 mm and 79.035 mm for the first two CPBSs, the power of the RHP mode is almost completely coupled into core B and the LHP mode remains in core A. Additionally, for the third CPBS, the RHP mode remains in core A, and the LHP mode is fully coupled into core B. Their ERs can respectively reach 160.73 dB, 140.31 dB and 120.00 dB at 1.55 μm, which are larger than those of most LPBSs and the dielectrically chiral DC-PCF CPBS [16,32] and are far more than of the structurally chiral DC-PCF CPBS [32], especially when the first ER is larger than 151.42 dB in the DC-PCF LPBS [16]. Additionally, the width of the spectra are, respectively, 22 nm, 12 nm and 12 nm when ER > 20 dB, which is smaller than those of most LPBSs and the dielectrically chiral DC-PCF CPBS [32]; however, it is about twice that of the structurally chiral DC-PCF CPBS [6]. According to the above analysis, we can know the performance of CPBS 1 is better than that of the other two CPBSs because of the superhigh ER and the wider bandwidth of the spectrum. Furthermore, it demonstrates that the stronger coupling effect can decrease the spectral bandwidth.

In practical applications of the design, more microstructures can be moderately considered so that the chiral dual-core PCFs have zero or near-zero dispersion around the operation wavelength, so that no additional dispersion is introduced or so that it has a flat dispersion to obtain the wider spectral bandwidth. In addition, it is necessary to consider the mode field of the CPBS to match that of the applied fiber in its design. Otherwise, the loss will be large to decrease the performance of the optic circuit.

Table 1 lists the performance comparison between the proposed and the reported chiral DC-PCF CPBSs. According to Table 1, the CLs for this work are far less than 205.5 mm [32] and 24.76 mm [6], while the ERs are better than those in the previous work, especially the ER of CPBS 1, which can reach 160 dB, which is higher than that of the reported work. The bandwidth of the spectrum is narrower than 32 nm [32], while it is still wider than 11.43 [6], especially for CPBS 1, which can reach 22 nm, which is almost twice as much the structurally chiral DC-PCF CPBS [6]. The chirality strength decreases to 46°/mm, which is far less than that in the previous work [6,32]. Therefore, the performance of CPBS 1 is better than that of the other two CPBSs. By further optimizing the fiber's structure or/and enhancing the chirality strength of the medium, the CL would further shorten and the performance would further improve.

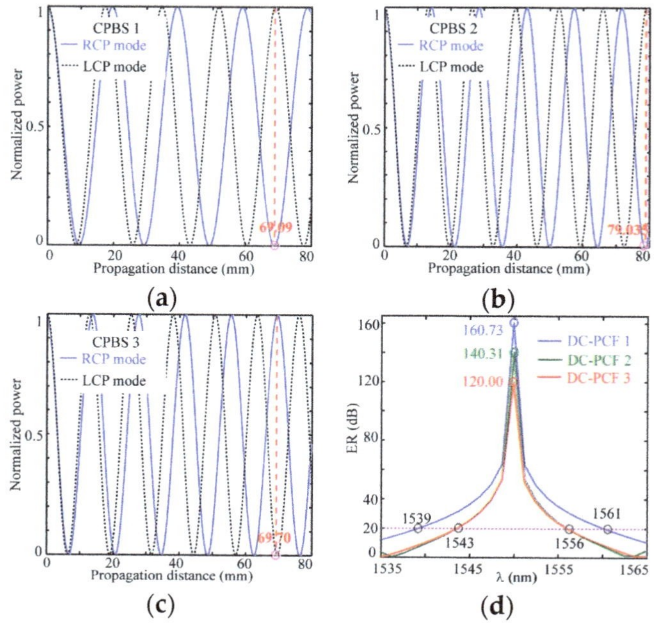

Figure 8. Normalized power in port A along propagation distance in the chiral PCF CPBSs in (**a**–**c**) and the corresponding dependence of the ER on wavelength in (**d**).

Table 1. Performance comparison between the proposed chiral DC-PCF CPBS and the reported chiral DC-PCF CPBS.

References		Working Wavelength	CL(mm)	ER	Spectral Bandwidth (nm) (>20 dB)	Specific Rotatory Power or Twist Rate (°/mm)
Suga-Restrepo et al. [6]		1.55 μm	24.76	50	11.43	899.54
Li et al. [32]		1.55 μm	205.5	120	32	96.6
Our work	CPBS 1	1.55 μm	69.09	160.73	22	46
	CPBS 2	1.55 μm	79.035	140.31	13	46
	CPBS 3	1.55 μm	69.7	120.00	13	46

In addition to separating the CP beam with opposite handedness for same wavelengths, the CPBS can also split CP beams with different wavelengths. To ensure that the modes are CP, the structure and medium is employed at above three chiral DC-PCFs, and the dependence of the CL on wavelengths for the LHP mode is considered because the chirality strength is stronger in the shortwave region and because the CL of LHP mode is always shorter than that of the RHP mode for the same wavelength. Figure 9 shows the CLs of the LHP modes with wavelengths for the three chiral DC-PCFs corresponding to Figure 8, which covers the second and third communication window.

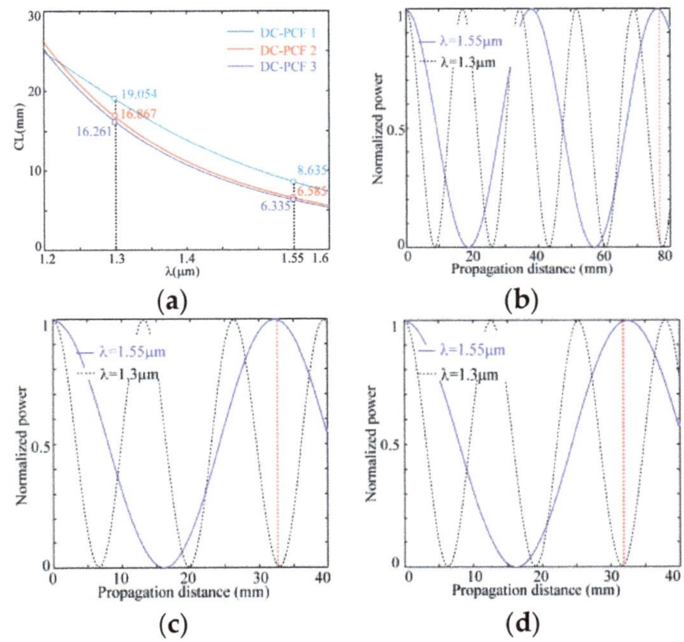

Figure 9. The variation relation of CLs with wavelengths on the LHP modes for the three DC-PCFs corresponding to Figure 8 in (**a**) and the normalized power variation along the propagation distance in core A in the three DC-PCFs for λ = 1.55 μm and λ = 1.30 μm in (**b**–**d**).

The coupling length also increases by decreasing the operating wavelength with the DC-PCF LPBS [1], where the CLs of the LHP modes are, respectively, 8.635 mm, 6.585 mm and 6.335 mm at λ = 1.55 μm and 19.054 mm, 16.867 mm and 16.261 mm at λ = 1.30 μm, and their corresponding ratios are 2.207, 2.561 and 2.460 when approaching 9/4, 5/2, 5/2, which meet the requirement for $CL_{\lambda 1}/CL_{\lambda 2}$ = odd/even or $CL_{\lambda 1}/CL_{\lambda 2}$ = even/odd [1]. This suggests that these fibers can be designed as a multiplexer/demultiplexer for separating the CP beams that approach λ = 1.55 μm and λ = 1.30 μm applied in fiber laser communication systems, and the corresponding physical lengths are, respectively, close to 76.216 mm, 33.734 mm and 32.522 mm, as shown in Figure 9b. The length of the multiplexer/demultiplexer can be made shorter through the further optimization of the structure of the fiber and the chirality of the medium.

4. Conclusions

A kind of efficient CPBS based on chiral DC-PCF was presented and analyzed by using the chiral PWE method, where the shorter distance between the two cores was selected. Based on the competitive effect in CL and the polarization between the circular asymmetry and the chirality of the medium, through the optimization of the structure, the DC-PCF CPBS was obtained for the central wavelength of 1.55 μm. The required chirality strength can decrease to $\delta = 46^{\circ}/\mathrm{mm}$, which is far less than $\delta = 96^{\circ}/\mathrm{mm}$; the CL can shorten to 69.09 mm, which is far less than the previous 205 mm; the width of spectra can reach 22 nm, which can be a little bit narrower than 32 nm when $|ER| > 20$ dB; and the ER above 160 dB is higher than 120 dB [32]. This performance of the CPBS was also better than that of the structural CPBS, except that the CL is longer [6]. Furthermore, on the basis of different wavelengths, the chiral DC-PCFs were discussed as multiplexers/demultiplexers. The simulation demonstrated that the multiplexer/demultiplexer can be perfectly realized for

CP beams around 1.55 μm/1.30 μm by further optimizing the structure of the fiber, which has potential applications in circular polarization laser communication systems.

The high-performance circular polarization beam splitter (CPBS) needs a medium with a stronger chirality to tailor the coupling length (CL), extinction ratio (ER) and circular polarization–maintaining characteristics. The chiral medium can be obtained through organic chemical synthesis [30,31] and the solution-doping technique [43]. Compared with the structurally chiral CPBS, the chiral dual-core PCF (DC-PCF) can be fabricated as the ordinary achiral one, such as by using the stack-and-draw technique, the drilling method, and so on. With the development of chiral materials and through the further optimization of the fiber structure, the performance of the dielectrically chiral DC-PCF CPBS can be further improved.

The CPBS can be applied in the field of sensors and optical fiber communication as a circular polarization–maintaining device to replace the existing LPBS and quarter-wave plates, to reduce the polarization error/noise and to oppositely split CP light and CP multiplexer/demultiplexer in fiber laser communication systems.

Author Contributions: Conceptualization, S.L. and Y.L.; methodology, S.L.; software, C.J.; validation, S.L.; formal analysis, H.G. and C.J.; investigation, S.L. and H.L.; resources, Y.L. and H.L.; data curation, S.L. and Y.L.; writing—original draft preparation, S.L.; writing—review and editing, S.L.; visualization, Q.S.; supervision, Y.L. and S.L.; project administration, S.L. and Q.S.; funding acquisition, Y.L. All authors have read and agreed to the published version of the manuscript.

Funding: This research was funded by the Fundamental Research Funds for the Central Universities under grant no. 3072022CF0806.

Institutional Review Board Statement: Not applicable.

Informed Consent Statement: Not applicable.

Data Availability Statement: Not applicable.

Conflicts of Interest: The authors declare no conflict of interest.

References

1. Saitoh, K.; Sato, Y.; Koshiba, M. Coupling characteristics of dual-core photonic crystal fiber couplers. *Opt. Express* **2003**, *11*, 3188–3195. [CrossRef]
2. Zhang, L.; Yang, C. Polarization splitter based on photonic crystal fibers. *Opt. Express* **2003**, *11*, 1015–1020. [CrossRef] [PubMed]
3. Lin, Z.; Yang, C.; Yu, C.; Luo, T.; Alan, E. PCF-based polarization splitters with simplified structures. *J. Light. Technol.* **2005**, *23*, 3558–3565.
4. Hu, D.; Ping, S.; Chao, L.; Sun, X.; Ren, G.; Xia, Y.; Wang, G. Design and analysis of thermally tunable liquid crystal filled hybrid photonic crystal fiber coupler. *Opt. Commun.* **2009**, *282*, 2343–2347. [CrossRef]
5. Zhang, S.; Yu, X.; Zhang, Y.; Shum, P.; Zhang, Y.; Xia, L.; Liu, D. Theoretical study of dual-core photonic crystal fibers with metal wire. *IEEE Photonics J.* **2012**, *4*, 1178–1187. [CrossRef]
6. Suga-Restrepo, J.E.; Guimares, W.M.; Franco, M. All-fiber circular polarization beam splitter based on helically twisted twin-core photonic crystal fiber coupler. *Opt. Fiber Technol.* **2020**, *58*, 102285. [CrossRef]
7. Jiang, H.; Wang, E.; Zhang, J.; Lei, H.; Mao, Q.; Li, Q.; Xie, K. Polarization splitter based on dual-core photonic crystal fiber. *Opt. Express* **2014**, *22*, 30461–30466. [CrossRef] [PubMed]
8. Xu, Z.; Li, X.; Ling, W.; Liu, P.; Zhang, Z. Design of short polarization splitter based on dual-core photonic crystal fiber with ultra-high extinction ratio. *Opt. Commun.* **2015**, *354*, 314–320. [CrossRef]
9. Sheng, Z.; Wang, J.; Feng, R. Design of a compact polarization splitter based on the dual-elliptical-core photonic crystal fiber. *Infrared Phys. Technol.* **2014**, *67*, 560–565. [CrossRef]
10. Zou, H.; Xiong, H.; Zhang, Y.S.; Ma, Y.; Zheng, J.J. Ultra-broadband polarization splitter based on graphene layer-filled dual-core photonic crystal fiber. *Chin. Phys. B* **2017**, *26*, 124216. [CrossRef]
11. Zi, J.; Li, S.; Wang, G.; An, G.; Fan, Z. Design of ultra-short polarization beam splitter based on liquid-filled photonic crystal fiber. *Opt. Quantum Electron.* **2016**, *48*, 233. [CrossRef]
12. Dou, C.; Jing, X.; Li, S.; Wu, J.; Wang, Q. A compact and low-loss polarization splitter based on dual-core photonic crystal fiber. *Opt. Quantum Electron.* **2018**, *50*, 255. [CrossRef]

13. Xu, Q.; Zhao, Y.; Xia, H.; Lin, S.; Zhang, Y. Ultrashort polarization splitter on dual-core photonic crystal fibers with gold wire. *Opt. Eng.* **2018**, *57*, 46104. [CrossRef]
14. Younis, B.M.; Hameed, M.F.O.; Obayya, S.S.A. Tunable polarization splitter based on asymmetric dual-core liquid photonic crystal fiber. *Opt. Quantum Electron.* **2021**, *53*, 51. [CrossRef]
15. Chu, L.; Liu, M.; Shum, P.; Fu, Y. Simultaneous achievement of an ultrashort length and a high extinction ratio polarization splitter based on the dual-core photonic crystal fiber with Ge20Sb15Se65 glass. *Appl. Opt.* **2019**, *58*, 7892–7896. [CrossRef] [PubMed]
16. Lou, J.; Yang, Y.; Zhang, X.; Qu, Q.; Li, S. Ultra-Short Dual-Core Photonic Crystal Fiber Polarization Beam Splitter with Round Lattice and As_2S_3-Filled Center Air Hole. *Photonics* **2022**, *9*, 36. [CrossRef]
17. Sanders, G.A.; Demma, N.; Rouse, G.F.; Smith, R.B. Evaluation of polarization maintaining fiber resonator for rotation sensing applications. In *Optical Fiber Sensors*; Optica Publishing Group: Washington, DC, USA, 1988; p. FBB7.
18. Wang, X.; He, Z.; Hotate, K. Polarization-noise suppression by twice 90-degree polarization-axis rotated splicing in resonator fiber optic gyroscope. In Proceedings of the Conference on Lasers & Electro-Optics, Baltimore, MD, USA, 2–4 June 2009.
19. Wang, Z.; Wang, G.; Wang, Z.; Gao, W.; Cheng, Y. Incoherence suppression method of optical noises in resonant fiber optic gyro based on the circularly polarized light propagation mechanism in resonator. *Opt. Lett.* **2021**, *46*, 3191–3194. [CrossRef]
20. Gao, H.; Wang, G.; Gao, W.; Li, S. A chiral photonic crystal fiber sensing coil for decreasing the polarization error in a fiber optic current sensor. *Opt. Commun.* **2020**, *469*, 125755. [CrossRef]
21. Gao, W.; Gao, H.; Xia, X.; Wang, G.; Li, S.; Zhao, Y.; Wang, Y. Design of a dielectric chiral micro-structured fiber applied in a fiber optic current sensor. *Opt. Contin.* **2022**, *1*, 271–282. [CrossRef]
22. Tentroi, D.; Treviño-Martínez, F.; Ayala-Díaz, C.; Mendieta-Jiménez, F.J.; Tentori, D. Birefringence assessment of single-mode optical fibers. *Opt Express* **2005**, *13*, 2556–2563.
23. Kopp, V.I.; Park, J.; Wlodawski, M.; Singer, J.; Neugroschl, D.; Genack, A.Z. Chiral Fibers: Microformed Optical Waveguides for Polarization Control, Sensing, Coupling, Amplification, and Switching. *J. Light. Technol.* **2014**, *32*, 605–613. [CrossRef]
24. Kopp, V.I.; Genack, A.Z. Double-helix chiral fibers. *Opt. Lett.* **2003**, *28*, 1876–1878. [CrossRef] [PubMed]
25. Kopp, V.I.; Churikov, V.M.; Singer, J.; Chao, N.; Neugroschl, D.; Genack, A.Z. Chiral fiber gratings. *Science* **2004**, *305*, 74–75. [CrossRef] [PubMed]
26. Peng, N.; Huang, Y.; Wang, S.B.; Wen, T.; Liu, W.; Zuo, Q.; Wang, L. Fiber optic current sensor based on special spun highly birefringent fiber. *IEEE Photonics Technol. Lett.* **2013**, *25*, 1668–1671. [CrossRef]
27. Li, J.Q.; Su, Q.Y.; Cao, Y.S. Circularly polarized guided modes in dielectrically chiral photonic crystal fiber. *Opt. Lett.* **2010**, *35*, 2720–2722. [CrossRef]
28. Li, S.; Li, J.Q. Chiral photonic crystal fibers with single mode and single polarization. *Opt. Commun.* **2015**, *356*, 96–102. [CrossRef]
29. Li, S.; Li, J.Q.; Cao, Y.S. Guided modes in squeezed chiral microstructured fibers. *Opt. Commun.* **2015**, *341*, 79–84. [CrossRef]
30. Argyros, A.; Straton, M.; Docherty, A.; Min, E.H.; Ge, Z.; Wong, K.H.; Ladouceur, F.; Poladian, L. Consideration of chiral optical fibres. *Front. Optoelectron. China* **2010**, *3*, 67–70. [CrossRef]
31. Min, E.H.; Wong, K.H.; Setijadi, E.; Ladouceur, F.; Straton, M.; Argyros, A. Menthol-based chiral copolymers for polymer optical fibres (POF). *Polym. Chem.* **2011**, *2*, 2045–2051. [CrossRef]
32. Li, S.; Li, Y.; Sun, Q.; Gao, H.; Ji, C.; Lv, H. Circular-polarized beam splitter Based on dielectrically chiral dual-core photonic crystal fiber. *Opt. Contin.* **2022**, *1*, 2442–2552. [CrossRef]
33. Gao, F.; Luo, S.; Ji, H.M.; Yang, X.G.; Liang, P.; Yang, T. Broadband tunable InAs/InP quantum dot external-cavity laser emitting around 1.55 μm. *Opt. Express* **2015**, *23*, 18493–18500. [CrossRef] [PubMed]
34. Jopson, R.M.; Eisenstein, G.; Whalen, M.S.; Hall, K.L.; Koren, U.; Simpson, J.R. A 1.55-μm semiconductor-optical fiber ring laser. *Appl. Phys. Lett.* **1986**, *48*, 204–206. [CrossRef]
35. Zheng, Z.; He, C.; Wen, Y.; Li, H. Performance Tunable Passively Q-switched Fiber Laser Based on Single Wall Carbon Nanotube. *Mod. Phys. Lett. B* **2022**, *36*, 2250054. [CrossRef]
36. Huffaker, D.L.; Park, G.; Zou, Z.; Shchekin, O.B.; Deppe, D.G. 1.3 μm room-temperature gaas-based quantum-dot laser. *Appl. Phys. Lett.* **1998**, *73*, 2564–2566. [CrossRef]
37. Liu, M.; Shi, J.; Ma, D.; Jin, E.; Li, S. 1.3 μm high-power superluminescent diode. *Semicond. Photonics Technol.* **1995**, *1*, 67–70.
38. Wen, Y.; He, C.; Zheng, J.; Fang, Y. Passively Q-switched erbium-doped fiber laser based on carbon nanotube saturable absorber. *Proc. SPIE* **2022**, *12478*, 1247810.
39. Ferrando, A.; Silvestre, E.; Miret, J.J.; Andrés, P. Nearly zero ultraflattened dispersion in photonic crystal fibers. *Opt. Lett.* **2000**, *25*, 790–792. [CrossRef] [PubMed]
40. Wang, W.; Hou, L.T.; Ming, L.; Zhou, G.Y. Design of double cladding nearly zero dispersion flattened nonlinear photonic crystal fiber. *Chin. Phys. Lett.* **2009**, *26*, 114205.
41. Liu, Z.; Hou, L.; Wang, W. Tailoring nonlinearity and dispersion of photonic crystal fibers using hybrid cladding. *Braz. J. Phys.* **2009**, *39*, 50–54.
42. Ishigure, T.; Nihei, E.; Koike, Y. Optimum refractive-index profile of the graded-index polymer optical fiber, toward gigabit data links. *Appl. Opt.* **1996**, *35*, 2048–2053. [CrossRef]

43. Large, M.; Ponrathnam, S.; Argyros, A.; Pujari, N.S.; Cox, F. Solution doping of microstructured polymer optical fibres. *Opt. Express* **2004**, *12*, 1966–1971. [CrossRef] [PubMed]
44. Pan, X.F.; Tao, W.D.; Yan, F.B.; Bai, G.R. Fabrication of PMMA doped with griseofulvin material and its optical rotatory dispersion. *Optoelectron. Lett.* **2006**, *2*, 41–43. [CrossRef]

photonics

MDPI

Article

Characterization of Grid Lines Formed by Laser-Induced Forward Transfer and Effect of Laser Fluence on the Silver Paste Transformation

Yucui Yu [1,2,3,4], Yanmei Zhang [1,2,3,4], Chongxin Tian [1,2,3], Xiuli He [1,2,3], Shaoxia Li [1,2,3,*] and Gang Yu [1,2,3,4,*]

1 Institute of Mechanics, Chinese Academy of Sciences, Beijing 100190, China; yuyucui@imech.ac.cn (Y.Y.); zhangyanmei2@imech.ac.cn (Y.Z.); tianchongxin@imech.ac.cn (C.T.); xlhe@imech.ac.cn (X.H.)
2 Guangdong Aerospace Research Academy (Nan Sha), Guangzhou 511458, China
3 School of Engineering Science, University of Chinese Academy of Sciences, Beijing 100049, China
4 Center of Materials Science and Optoelectronics Engineering, University of Chinese Academy of Sciences, Beijing 100049, China
* Correspondence: Correspondence: lisx@imech.ac.cn (S.L.); gyu@imech.ac.cn (G.Y.)

Abstract: The investigation of novel approaches for forming solar cell grid lines has gained importance with the rapid development of the photovoltaic industry. Laser-induced forward transfer (LIFT) is a very promising approach for microstructure fabrication. In this work, the morphology of grid lines deposited by LIFT was investigated. A characterization scheme for solar cell grid lines was proposed. The shape of grid lines was described, combined with confocal imaging. The evolution process of grid lines from no forming to single-peak and double-peak with a variation of laser fluence was observed. According to experimental conditions, different types of grid line morphology were obtained and transfer mechanisms of silver paste were proposed based on fluid dynamics. The influence of laser fluence on the morphology of formed grid lines was explained through phenomenology and analysis. This can provide a guide for morphology control in forming the process of grid lines.

Keywords: micro structure fabricating; LIFT; solar cell grid lines; morphological characteristics; printing mechanisms

Citation: Yu, Y.; Zhang, Y.; Tian, C.; He, X.; Li, S.; Yu, G. Characterization of Grid Lines Formed by Laser-Induced Forward Transfer and Effect of Laser Fluence on the Silver Paste Transformation. *Photonics* **2023**, *10*, 717. https://doi.org/10.3390/photonics10070717

Received: 28 April 2023
Revised: 15 June 2023
Accepted: 15 June 2023
Published: 22 June 2023

1. Introduction

Laser-induced forward transfer (LIFT) is one of the most promising approaches for contactless device fabrication at the microscale or nanoscale levels [1]. In classical LIFT, the laser beam irradiates a thin film of an absorbing material (the donor), which is usually deposited onto a transparent substrate. A pulsed laser is used to ablate a small portion of the donor film and deposit it onto a receiver substrate. The donor material can be solid or liquid and is transferred in the solid phase or as small droplets in the liquid phase onto the receiver substrate. The donor material can be a solid or liquid material, and is transferred in the solid phase or as small droplets in the liquid phase onto the receiver substrate. The flexibility, simplicity, and high speed of laser-induced microfabrication and nanofabrication techniques give them advantages over traditional device manufacturing techniques [2–4]. LIFT is used to print metal contacts [5] and patterned solder paste [6] for microwave interconnects in microelectronics or coplanar waveguides [7].

With the development of solar cell technology [8], reducing cost and increasing efficiency have become the goals of industry development. As an important part of the solar cell, a grid line is used to transmit the generated photogenerated carriers. Therefore, the formation of grid lines (metallization of cells) is a necessary process in the fabrication of solar cells. Additionally, the amount of silver paste used in the grid line is the second-largest cost for solar cell production. The morphological characteristics of grid lines greatly influence the performance of solar cells. To reduce cost and improve efficiency, the width of grid lines needs to be much thinner, and the aspect ratio needs to be much larger. This will reduce the amount of silver paste and the shielding of grid lines on the solar cell surface. Screen printing is usually used to form grid lines in the industry. The process is mature and the production rate is fast. However, as a kind of contact printing technology, screen printing inevitably exerts certain pressure on the silicon wafer during the printing process, which may cause the silicon wafer to break. In addition, the use of high-viscosity silver paste particles in front metallization will lead to screen blocking, affecting the quality of printing. Non-contact printing methods, such as aerosol jetting [9], inkjet [10], microprinting [11], and dispensing printing [12,13] are studied. As a very promising approach, LIFT is being tried for the front metallization of solar cells [14]. Benefiting from being non-contact and maskless, the damage to the fragile silicon wafer and accuracy reduction caused by the blockage of high-viscosity paste are avoided. Moreover, due to the high controllability of laser, the morphology of grid lines could be controlled more flexibly, and the electrical properties could be improved.

The electrical properties, such as conductivity and contact resistivity, of the grid line depend not only on the material properties themselves but also on the morphology of the grid line. Currently, parameters for the cross-sectional characterization of grid lines mainly focus on the cross-sectional area, shielding area, the effective width of the core [15], and aspect ratio. However, there is no uniform standard, and these parameters are not sufficient to fully characterize the morphology of grid lines. To more completely characterize the morphology of grid lines and more accurately judge forming quality, parameters such as average shading line width, conductivity line width, line height, cross-sectional area and aspect ratio, valley height, and peak separation of grid lines are proposed for the analysis of the influence of laser fluence.

The transfer process of silver paste and the morphology, as well as the quality of the formed grid line, will be strongly affected by the specific rheology of the silver paste used [16], making the transfer mechanism different from that previously observed in conductive inks or nano pastes [17]. Continuous lines with high aspect ratios have been printed at high speeds [18]; however, in this research, the influence of the interaction on adjacent pulses is not considered. The grid line obtained is wavy, which greatly affects the conduction efficiency. Both P.Sopeña [19] and Muñoz Martin [20] focused on the interaction between adjacent pulses. A non-uniform donor film surface was generated. Different events [21], such as bubble expansion, bubble burst, or no expansion at all, can randomly develop depending on the donor film state at the moment of irradiation, which causes transmission failure. To improve grid line quality and avoid non-uniformities or other defects, researchers have proposed many methods. Multiple line printing [22] was studied, which consists of printing several lines on top of the previous line. In this case, the donor was moved after each voxel transfer, and each pulse irradiated a new area without interfering with the others. Finally, a grid line with a width of 65 μm was obtained. However, the printing speed of this approach is slow due to the limitation of the donor's movement speed. Y. Chen [18] and Muñoz Martin [20] both increased the laser spot pitch to reduce the influence of the interaction between adjacent pulses. However, the increase in the spot pitch caused the overlapping area of voxels to shrink, which resulted in significant fluctuations in the line width. Muñoz Martin [20] solved this problem by using a higher laser fluence to obtain larger voxels, which could increase the overlapping area of voxels. Nevertheless, this method resulted in a significant increase in the width of the obtained line, reaching 134 μm. Pattern transfer printing (PTP) [23] is also a non-contact laser printing technique similar to LIFT. A transparent polymer with narrow grooves is applied as the transparent substrate in PTP, and silver paste can only be filled into the grooves. Laser beam will scan along the grooves and the silver paste is transferred from the grooves to the silicon wafer. The width and aspect ratio of the grid lines could be controlled by adjusting the size of the grooves. The latest research [24] shows that PTP could obtain grid lines with a width of 20 μm and an aspect ratio of 0.6. Compared with LIFT, PTP can obtain grid lines with a width of 20 μm and an aspect ratio of 0.6. Compared with LIFT, PTP has a higher printing resolution. However, the transparent substrate must be customized to the printed pattern before transferring. Additionally, as processing time extends, drying silver paste may clog narrow grooves, leading to the invalidation of this method. Thus, it remains a big challenge to print high-resolution grid lines with the merits of non-contact and non-masking methods. To avoid interaction and achieve successful transfer, a larger distance between laser points was used. However, in cases of large spot spacing, it is difficult to obtain continuous grid lines or even thinner lines. Therefore, analyzing transfer mechanisms at high repetition frequencies for forming grid lines is very important. In this paper, the interaction between pulses is considered, and the grid line is formed under a small gap and high overlap rate with a high repetition frequency (1000 kHz).

In this paper, grid lines are formed by a non-contact laser forming method. Two typical morphologies of grid lines, namely single-peak, and double-peak, were observed using high-viscosity silver paste in the LIFT forming process. The typical morphology was characterized and the influences of laser fluence on the morphology of transferred paste and grid lines were studied. Based on characterization of grid lines and transfer analysis, the transfer mechanism of LIFT for high-viscosity silver paste was proposed in forming grid lines.

2. Materials and Methods

A self-built LIFT system was employed. As shown in Figure 1, a pulsed laser (532 nm; 12 ps; 1 MHz) was used as the laser source, a galvanometric mirror system as a scanning head, high precision fused quartz as the donor substrate, high-viscosity silver paste as the donor film, and monocrystalline silicon wafer as the receiving substrate. With a galvanometric mirror system, the system can scan at speeds up to 5 m/s along a predetermined path. An f-theta lens was applied to focus the laser beam reflected from the galvanometer scanner onto the donor. The distance from the donor to the receiving substrate could be electrically adjusted from 15~50 µm. The laser fluence can be calculated using Equation (1).

$$F_0 = \frac{E_0}{A} \tag{1}$$

where F_0 is the laser fluence. E_0 is the single pulse energy. A is the beam area. By adjusting the output power of the laser, the fluence can be controlled, when the laser spot size, speed and frequency rate are fixed. Different laser fluences (4.91 to 11.04 J/cm^2) were used to form grid lines.

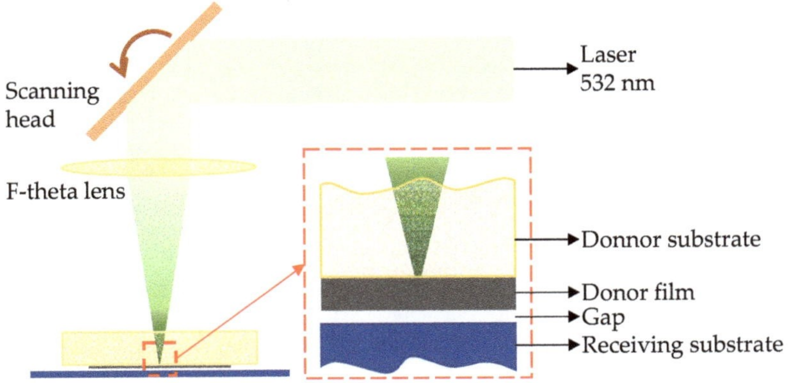

Figure 1. LIFT schematic diagram.

The parameters of the silver paste used as the donor material for LIFT are shown in Table 1. With a high-viscosity paste of 157 Pa·s at a shear rate of 20 rpm/3 min, the silver paste exhibits non-Newtonian thixotropic fluid with pseudoplastic (viscosity decreases with increasing shear rate until an equilibrium value is reached) and thixotropic (viscosity at a constant shear rate decreases and eventually steadies out to an equilibrium value) behavior. In other words, thixotropy is the thinning of shear force during usage and a rapid rebound to the initial structure after force is no longer applied. Relevant parameters of silver paste were obtained through HAAKE Rheostress 6000 with shear rates from 0.01 to 1000 s^{-1}, at 25 °C. The viscosity curve and flow curve present basic information on paste rheological properties. From the viscosity curve, it can be seen that when fluid is subjected to shear strain, its viscosity decreases with increasing shear rate. From the flow curve, shear stress increases with increasing shear rate, meaning silver paste has the characteristic of shear thinning. The effect of shear thinning is to reduce the flow resistance and friction to improve the fluidity and lubricity of the fluid. The thickness of the paste layer on the donor substrate is 16.9 μm, as shown in Figure 2. As a kind of non-Newtonian fluid, silver paste was stirred for approximately 10–15 min to attain equilibrium viscosity before use. To avoid viscosity changes caused by the evaporation of liquid components from the film, printing should be completed within 10 min after sample preparation. Otherwise, a change in viscosity will seriously affect transfer results.

Table 1. Sliver paste properties.

Properties	Value
Solid content	91.68 wt%
Particle size	<3~5 μm
Viscosity	157 pa·s(20 rpm/3 min)

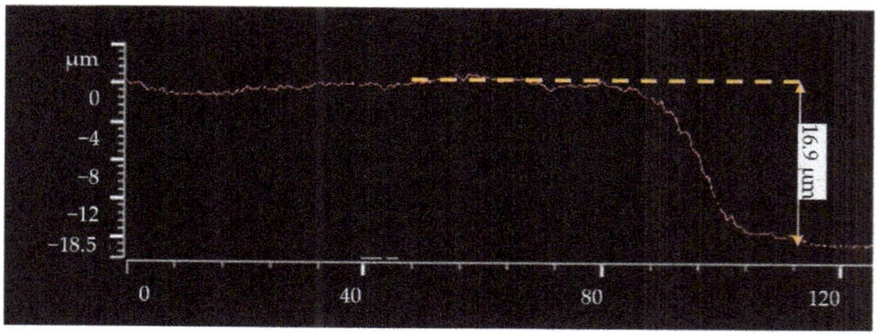

Figure 2. Thickness of paste layer on the donor substrate measured by laser scanning confocal microscope.

The monocrystalline silicon wafer was chosen as the acceptor under the silver paste, with an average line roughness of 0.2 μm. The distance between the donor and acceptor could be adjusted from 0 μm to 100 μm. After experiments, to evaporate solvents and bond silver powders, silicon wafers were subsequently dried and sintered at 200 °C and 750 °C separately. The morphology of transferred lines was measured by an optical microscope and laser scanning confocal microscope Olympus LEXT OLS5100 (LSCM).

3. Results

3.1. Characterization

The conductivity of the grid line is greatly affected by its geometry. Therefore, it is very important to evaluate the 3D geometry of contacting grids with sufficient measurement methods and statistical evaluation. With LIFT, two typical morphologies of grid lines were formed, they have a cross section with a single-peak and double-peak, respectively. As shown in Figure 3, to ensure adequate statistical evaluation and repeatability, measurements of each x-axis are averaged to avoid subjective effects and measurement errors from manual measurements. The obtained data are also averaged across the y-axis to determine the wafer datum and avoid pyramidal effects.

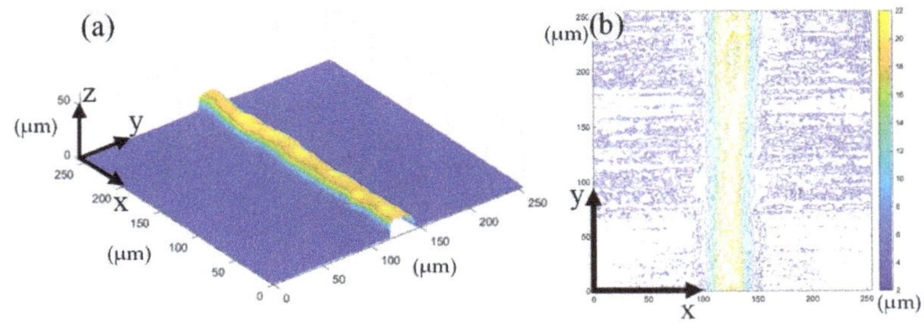

Figure 3. (**a**) The 3D geometry of the grid line measured by Laser Scanning Confocal Microscope; (**b**) Top view contour map of grid line measured by Laser Scanning Confocal Microscope.

Ideally, a grid line with a larger cross-sectional area, a larger height, and a smaller width is needed to maximize the optical and electrical properties of the grid line and increase using the efficiency of silver paste. A representation system was defined to evaluate grid line geometry from the cross-section direction. According to a grid line with single-peak and double-peak, respectively, the average shading line-width, conductivity line-width, line height, cross-sectional area and aspect ratio, valley height, and peak separation of grid lines are defined, as shown in Figure 4 and Table 2.

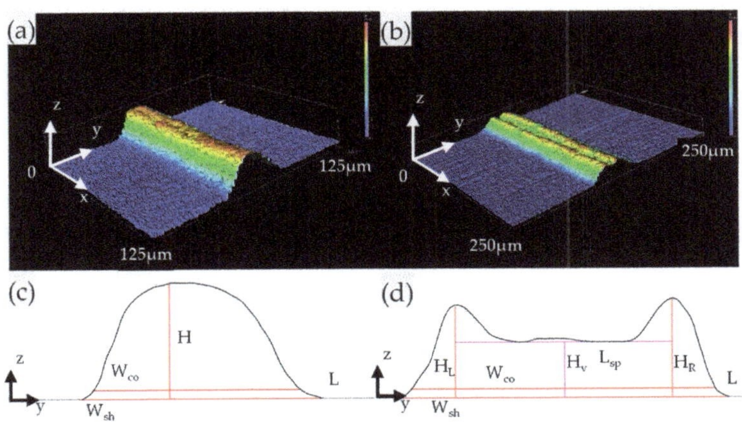

Figure 4. Morphology of grid lines with single-peak (**a**), and double-peak (**b**); Parameters defined for grid lines with single-peak (**c**), and double-peak (**d**).

Table 2. Morphologies of grid lines with single-peak and double-peak.

Morphologies	Single-Peak	Double-Peak
Horizontal baseline	L	L
Average shading line-width	W_{sh}	W_{sh}
Conducting line-width	W_{co}	W_{co}
Line height	H	H_L; H_R
Cross-sectional area	A_{crsct}	A_{crsct}
Aspect ratio	AR	AR
Valley height	/	H_v
Peak separation	/	L_{sp}

To separate silver paste drift from the amount of silver paste that contributes to the conductivity of the grid line, average shading width is defined as the length of the line segment at a height of 2 μm from the horizontal baseline (silicon wafer fluctuation ± 2 μm). From Figures 3b and 4a,b, it can be seen that when the relative height difference is 2 μm, there is a coherent grating edge (conductive grid edges). A horizontal baseline with an average length of 50 μm on both sides of the measurement result is selected to eliminate the tilt error of the measurement. The average shading line-width is the intersection of the cross-section curve and horizontal baseline. The conducting line-width is 2 μm above the horizontal baseline. Line height is the maximum distance between the cross-section curve and the horizontal baseline. Line height is also the maximum distance of the left peak between cross-section curve and the horizontal baseline and maximum distance of the right peak between the cross-section curve and the horizontal baseline. Cross-sectional area is the area enclosed by the horizontal baseline and cross-sectional curve. Aspect ratio is the ratio of line height H to conductive line width W_{co}. Valley height is valley height at the fitted line, distance from the horizontal baseline. Peak separation is the horizontal distance between two highest peaks.

3.2. Effect of Laser Fluence on Cross-Sectional Morphology

Based on the above experimental system, a variety of grid lines with different morphologies were formed under different laser fluences. After laser irradiation, the donor was removed from the acceptor substrate, leaving transferred lines. Grid line morphology at different laser fluences measured by LSCM is shown in Figure 5. When using the lowest fluence, there is no transfer, as shown in Figure 5a, suggesting the existence of a laser fluence transfer threshold. When laser fluence increases, paste transfer occurs, with wider lines at higher laser fluences. Subgraphs highlighted in the red solid line box are 3D profiles of grid lines with a single-peak, as shown in Figure 5b–d. Subgraphs highlighted in red dotted frames are 3D profiles of grid lines with a double-peak, as shown in Figure 5e–i. There are several grid lines with different morphologies, and only single-peak grid lines exhibit characteristics of a high aspect ratio.

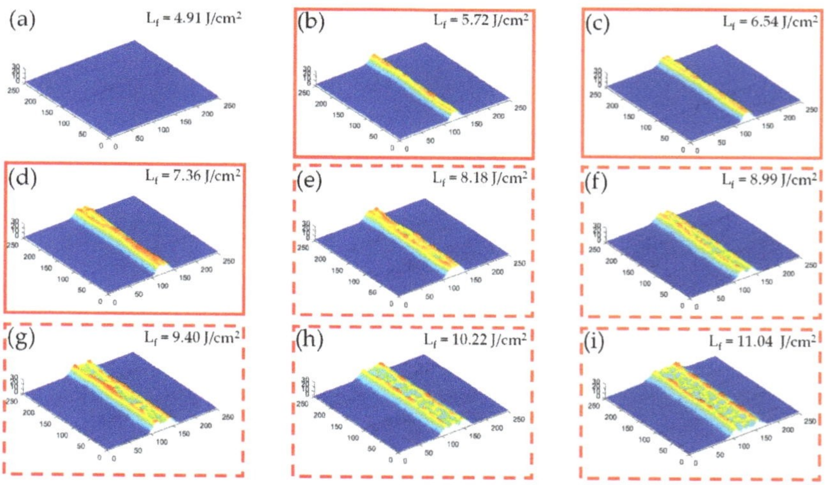

Figure 5. Grid line morphology at different laser fluences measured by LSCM, from (**a–i**), with the increase of the laser fluence.

Under different laser fluences, the cross-sectional morphologies of grid lines are significantly different, as shown in Figure 6. With increasing laser fluence, line width increases, and the distance between two peaks also increases. However, no matter the grid line with single-peak or double-peak, the heights of the peaks remain basically the same.

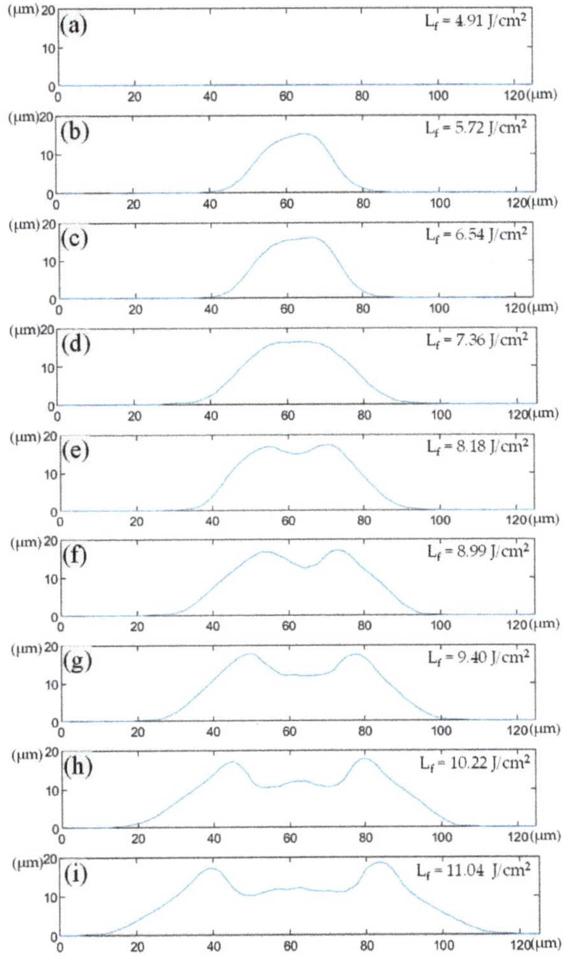

Figure 6. Contour of cross-sectional morphologies, from (**a–i**), with the increase of the laser fluence.

Figure 7 shows the results of transferred lines under different laser fluences. Specifically, the effect of laser fluence on line width, cross-sectional area, distance between two peaks, and line height is studied. Conductive line width, cross-sectional area, and double-peak distance increase with increasing laser fluence, and the increasing trend is essentially approximately linear, as shown in Figures 7a–c and 8. For single-peak lines, the default distance between two peaks is zero. The effect of laser fluence on height of the grid line is also shown in Figure 7d.

With changes in laser fluence, the height of the grid line can be considered basically the same, as shown in Figure 8. Line height is basically consistent with the thickness of the paste layer on the donor substrate (16.9 μm), as shown in Figure 2.

Therefore, it is reasonable to assume that there is only a vertical overall movement of paste at the peak but no horizontal flow during silver paste transfer. Based on the results shown in Figure 8, with increasing laser fluence, the aspect ratio shows a downward trend.

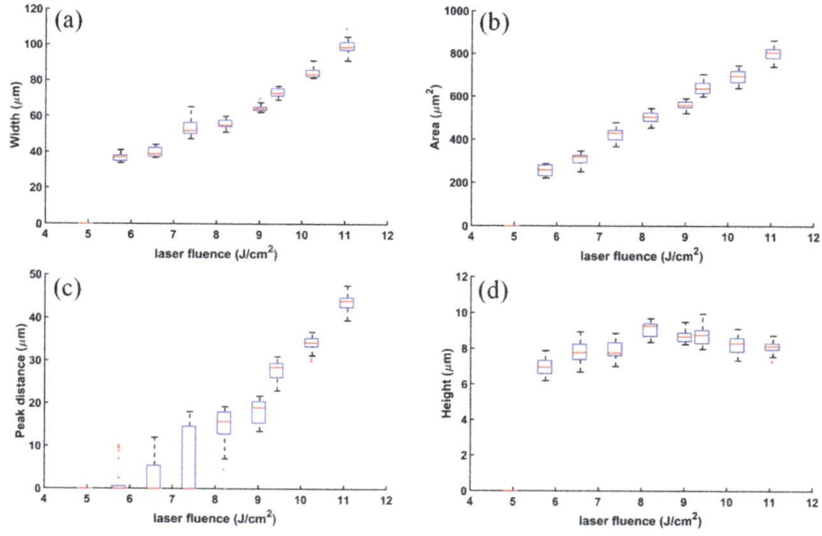

Figure 7. (**a**) The effect of laser fluence on width; (**b**) The effect of laser fluence on cross-sectional area; (**c**) The effect of laser fluence on distance between two peaks; (**d**) The effect of laser fluence on height.

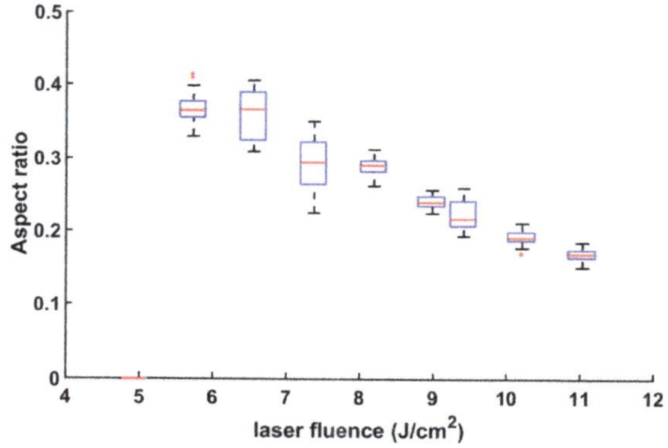

Figure 8. Effect of the laser fluence on aspect ratio of the grid line.

4. Discussion

We assume that the laser fluence is deposited at the interface between the donor substrate and that the silver paste film and the laser-induced plasma is generated. Plasma induces the shock wave propagating inside.

It is noted that the maximum pressure is independent of the pulse duration and is proportional to the square root of the laser fluence density [25]. Therefore, it is reasonable to analyze the evolution of the grid line morphology with the laser fluence.

When laser fluence just exceeds the transfer threshold, silver paste flows and bubbles form and develop until they come into contact with the silicon wafer. The silver paste and silicon wafer are separated and the grid line is formed, as shown in Figures 9 and 10. In Figure 9, H_s is the thickness of silver paste, D is the gap between silver paste and receiving substrate, R_b is bubble radius, H_b is bubble expansion height, and H_p is peak height. It is speculated that there is no transverse flow of silver paste when a single-peak appears. When laser fluence is larger, a portion of laser fluence is used to compress silver paste, which flows vertically and develops into peaks and valleys as laser fluence increases. Interaction between adjacent pulses is essentially a diffusion process between high-pressure bubbles and cavities. It usually takes about 20 μs from the expansion of high-pressure bubbles to the formation of cavities. A repetition rate as high as 1000 kHz means that the laser pulse interval is 1 μs and adjacent pulses interact with each other. A cavity is formed in the next laser pulse irradiation and the previous pulse irradiation area. The influence of pulse interaction is reduced by the constraint of a small gap. To summarize, experimental results on transferred lines show that three different regimes can be considered depending on laser fluence: no transfer, single-peak transfer, and double-peak transfer.

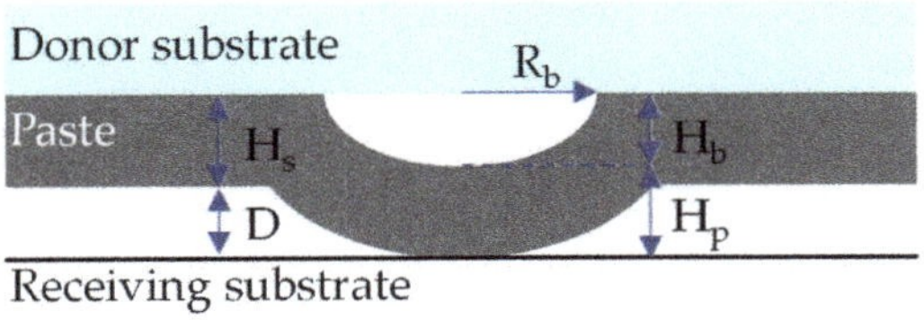

Figure 9. Bubble generation by laser induced.

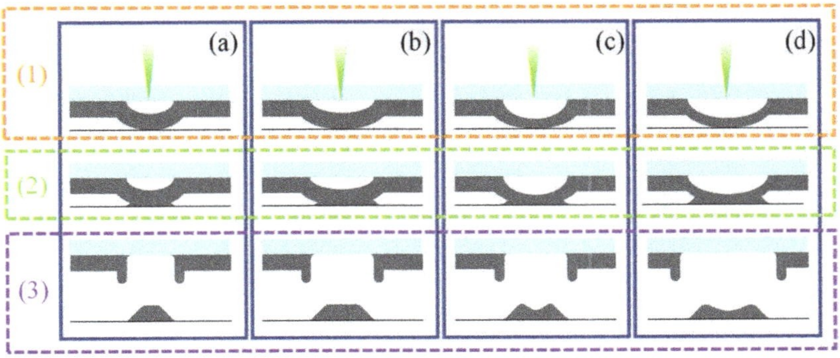

Figure 10. Forming process of line with single-peak (**a**,**b**), line with double-peak (**c**,**d**). (1) Silver paste foams and expands under pressure. (2) Silver paste contacts the receptor substrate and remains closed. (3) An external force causes the bubble to rupture and separate from the receptor substrate.

Before laser action, paste is almost unstressed and basically in a static state. First, the incident laser pulse is focused on the interface between glass and paste, rapidly heating and evaporating organic components of the silver paste. Molecules or particles in paste are rearranged or oriented under the action of shear force. A high-pressure and high-temperature vapor bubble is formed, as shown in Figure 10(1). There is a corresponding material transfer threshold for a given paste layer thickness. The laser fluence threshold of silver paste is attributed to the balance of internal and external forces. When paste is subjected to shear strain, its viscosity decreases with increasing shear rate. Flow resistance and friction of paste reduce, and both fluidity and lubricity of paste improve, manifested as bubble expansion. Then, due to the small gap, silver paste is pushed by the bubble to contact the receiving silicon wafer and continues to expand due to wettability. However, the bubble remains closed, and its original height is maintained after contacting the receptor, as shown in Figure 10(2). Finally, the paste is no longer stressed and tends to rebound to its initial structure until an external force is broken. Due to the thixotropy of the non-Newtonian fluid of the silver paste, the existing shape is maintained after stabilization. The silver paste at the joint is pulled off by a vertical upward pull, and the grid line is formed after separation, as shown in Figure 10(3).

As shown in Figure 10a, silver paste expands, contacts and infiltrates the receiving substrate, and finally the grid line forms. The vertical direction of the silver paste is not squeezed, so the peak height is the same as the thickness of the silver paste, which means $H_P = H_S$. As shown in Figure 10b, as laser fluence increases, a larger internal pressure overcomes ambient pressure and surface tension, and the bubble expands to a larger size. Silver paste is still not squeezed in the vertical direction, maintains its original height after contacting the acceptor, and expands horizontally along the substrate, showing an increase in line width. As shown in Figure 10c, with a further increase of laser fluence, the silver paste in the laser-action center is squeezed in the vertical direction, which results in local transverse flow, manifested as a decrease in the thickness in the middle position, that is, the valley between two peaks. However, the bubble size does not change much, and contact with the receiving substrate is not affected, and the original line width is still maintained. As laser fluence further increases, bubble size increases, and the contact area of the substrate increases too, extrusion becomes more obvious, and line width becomes bigger. With further increases in laser fluence, the transverse flow area becomes larger, showing a larger peak spacing, as shown in Figure 10d.

5. Conclusions

In this work, laser-induced forward transfer (LIFT) of high-viscosity silver paste was studied using a picosecond pulsed laser. Continuous grid lines with a high aspect ratio were formed, and 3D morphologies were measured by LSCM. It was found that there were two typical morphologies of grid lines: single-peak cross-section and double-peak cross-section. In order to better characterize the obtained grid lines for two typical morphologies, parameters such as shielding line width, effective line width, area, and height of the grid line were proposed and defined. The transfer mechanism was proposed, and the influence of laser fluence on paste flow was analyzed. Transfer modes according to the two typical morphologies were observed and defined. For single-peak transfer mode, when laser fluence just exceeds the transfer threshold, silver paste flows, then the bubble generates and develops until it makes contact with the silicon wafer and separates from the donor film, then forming a single-peak grid line. When laser fluence is relatively large, part of the laser fluence is used for silver paste compression and vertical flow. It develops into a peak valley as laser fluence increases, which results in a double-peak transfer mode. Research on the forming process could be a guide for future studies on morphology control in the forming process of grid lines and its application in solar cells.

Author Contributions: Conceptualization, Y.Y., C.T., Y.Z., S.L., X.H. and G.Y.; methodology, Y.Y., C.T., Y.Z., S.L., X.H. and G.Y.; software, Y.Y.; validation, S.L. and G.Y.; formal analysis, Y.Y.; investigation, Y.Y., C.T., Y.Z. and S.L.; resources, C.T.; data curation, Y.Y., C.T. and Y.Z.; writing original draft, Y.Y.; writing—review & editing, Y.Y. and S.L.; visualization, Y.Y.; supervision, S.L., X.H. and G.Y.; project administration, S.L. and G.Y.; funding acquisition, S.L. and G.Y. All authors have read and agreed to the published version of the manuscript.

Funding: This research was funded by High-Level Innovation Research Institute Program of Guangdong Province, grant number No. 2020 B0909010003 and Research Project of Guangdong Aerospace Research Academy, grant number GARA2022001000.

Institutional Review Board Statement: Not applicable.

Informed Consent Statement: Not applicable.

Data Availability Statement: The data presented in this study are available upon reasonable request.

Conflicts of Interest: The authors declare no conflict of interest.

References

1. Rehman, Z.U.; Yang, F.; Wang, M.; Zhu, T. Fundamentals and Advances in Laser-Induced Transfer. *Opt. Laser Technol.* **2023**, *160*, 109065. [CrossRef]
2. del Campo, A.; Arzt, E. Fabrication Approaches for Generating Complex Micro- and Nanopatterns on Polymeric Surfaces. *Chem. Rev.* **2008**, *108*, 911–945. [CrossRef] [PubMed]
3. Zhang, Y.; Liu, C.; Whalley, D. Direct-Write Techniques for Maskless Production of Microelectronics: A Review of Current State-of-the-Art Technologies. In Proceedings of the 2009 International Conference on Electronic Packaging Technology & High Density Packaging, Beijing, China, 10–13 August 2009; pp. 497–503.

4. Vaezi, M.; Seitz, H.; Yang, S. A review on 3D micro-additive manufacturing technologies. *Int. J. Adv. Manuf. Technol.* **2013**, *67*, 1721–1754. [CrossRef]
5. Florian, C.; Caballero-Lucas, F.; Fernández-Pradas, J.; Artigas, R.; Ogier, S.; Karnakis, D.; Serra, P. Conductive silver ink printing through the laser-induced forward transfer technique. *Appl. Surf. Sci.* **2015**, *336*, 304–308. [CrossRef]
6. Mathews, S.A.; Charipar, N.A.; Auyeung, R.C.Y.; Kim, H.; Piqué, A. Laser Forward Transfer of Solder Paste for Microelectronics Fabrication. In Proceedings of the Photonics West—Lasers and Applications in Science and Engineering, San Francisco, CA, USA, 7–12 February 2015.
7. Breckenfeld, E.; Kim, H.; Auyeung, R.; Charipar, N.; Serra, P.; Piqué, A. Laser-induced forward transfer of silver nanopaste for microwave interconnects. *Appl. Surf. Sci.* **2015**, *331*, 254–261. [CrossRef]
8. Acciari, G.; Adamo, G.; Ala, G.; Busacca, A.; Caruso, M.; Giglia, G.; Imburgia, A.; Livreri, P.; Miceli, R.; Parisi, A.; et al. Experimental Investigation on the Performances of Innovative PV Vertical Structures. *Photonics* **2019**, *6*, 86. [CrossRef]
9. Hörteis, M.; Glunz, S. Fine line printed silicon solar cells exceeding 20% efficiency. *Prog. Photovolt. Res. Appl.* **2008**, *16*, 555–560. [CrossRef]
10. Gizachew, Y.; Escoubas, L.; Simon, J.; Pasquinelli, M.; Loiret, J.; Leguen, P.; Jimeno, J.; Martin, J.; Apraiz, A.; Aguerre, J. Towards ink-jet printed fine line front side metallization of crystalline silicon solar cells. *Sol. Energy Mater. Sol. Cells* **2011**, *95*, S70–S82. [CrossRef]
11. Li, Q.; Grojo, D.; Alloncle, A.-P.; Chichkov, B.; Delaporte, P. Digital laser micro- and nanoprinting. *Nanophotonics* **2019**, *8*, 27–44. [CrossRef]
12. Pospischil, M.; Kuchler, M.; Klawitter, M.; Lacmago, I.; Tepner, S.; Efinger, R.; Linse, M.; Witt, D.; Gutscher, S.; Brand, A.; et al. High Speed Dispensing—A High-Throughput Metallization Technology for >21% PERC Type Solar Cells. In Proceedings of the 32nd European Photovoltaic Solar Energy Conference and Exhibition, Munich, Germany, 20–24 June 2016; pp. 403–406. [CrossRef]
13. Pospischil, M.; Riebe, T.; Jimenez, A.; Kuchler, M.; Tepner, S.; Geipel, T.; Ourinson, D.; Fellmeth, T.; Breitenbücher, M.; Buck, T.; et al. Applications of parallel dispensing in PV metallization. *AIP Conf. Proc.* **2019**, *2156*, 020005. [CrossRef]
14. Morales, M.; Chen, Y.; Muñoz, D.; Lauzurica, S.; Molpeceres, C. High volume transfer of high viscosity silver pastes using laser direct-write processing for screen printing of c-Si cells. In Proceedings of the Laser-based Micro- and Nanoprocessing IX, San Francisco, CA, USA, 1 March 2015; Volume 9351, p. 93510B. [CrossRef]
15. Strauch, T.; Demant, M.; Lorenz, A.; Haunschild, J.; Rein, S. Two Image Processing Tools to Analyse Alkaline Texture and Contact Finger Geometry in Microscope Images. In Proceedings of the 29th European Photovoltaic Solar Energy Conference and Exhibition, Amsterdam, The Netherlands, 23–25 September 2014; pp. 1132–1137. [CrossRef]
16. Zhang, Y.; Tian, C.; Yu, Y.; He, X.; Bian, Y.; Li, S.; Yu, G. Morphological Characteristics and Printing Mechanisms of Grid Lines by Laser-Induced Forward Transfer. *Metals* **2022**, *12*, 2090. [CrossRef]
17. Munoz-Martin, D.; Brasz, C.; Chen, Y.; Morales, M.; Arnold, C.; Molpeceres, C. Laser-induced forward transfer of high-viscosity silver pastes. *Appl. Surf. Sci.* **2016**, *366*, 389–396. [CrossRef]
18. Chen, Y.; Munoz-Martin, D.; Morales, M.; Molpeceres, C.; Sánchez-Cortezon, E.; Murillo-Gutierrez, J. Laser Induced Forward Transfer of High Viscosity Silver Paste for New Metallization Methods in Photovoltaic and Flexible Electronics Industry. *Phys. Procedia* **2016**, *83*, 204–210. [CrossRef]
19. Sopeña, P.; Fernández-Pradas, J.; Serra, P. Laser-induced forward transfer of conductive screen-printing inks. *Appl. Surf. Sci.* **2020**, *507*, 145047. [CrossRef]
20. Munoz-Martin, D.; Chen, Y.; Morales, M.; Molpeceres, C. Overlapping Limitations for ps-Pulsed LIFT Printing of High Viscosity Metallic Pastes. *Metals* **2020**, *10*, 168. [CrossRef]
21. Unger, C.; Gruene, M.; Koch, L.; Koch, J.; Chichkov, B. Time-resolved imaging of hydrogel printing via laser-induced forward transfer. *Appl. Phys. A* **2011**, *103*, 271–277. [CrossRef]
22. Sanchez-Aniorte, M.I.; Mouhamadou, B.; Alloncle, A.P.; Sarnet, T.; Delaporte, P. Laser-induced forward transfer for improving fine-line metallization in photovoltaic applications. *Appl. Phys. Mater. Sci. Process.* **2016**, *122*, 595. [CrossRef]
23. Lossen, J.; Matusovsky, M.; Noy, A.; Maier, C.; Bähr, M. Pattern Transfer Printing (PTPTM) for c-Si Solar Cell Metallization. *Energy Procedia* **2015**, *67*, 156–162. [CrossRef]
24. Adrian, A.; Rudolph, D.; Willenbacher, N.; Lossen, J. Finger Metallization Using Pattern Transfer Printing Technology for c-Si Solar Cell. *IEEE J. Photovolt.* **2020**, *10*, 1290–1298. [CrossRef]
25. Fabbro, R.; Fournier, J.; Ballard, P.; Devaux, D.; Virmont, J. Physical study of laser-produced plasma in confined geometry. *J. Appl. Phys.* **1990**, *68*, 775–784. [CrossRef]

photonics

MDPI

Article

Study on the Technology and Mechanism of Cleaning Architectural Aluminum Formwork for Concrete Pouring by High Energy and High Repetition Frequency Pulsed Laser

Kun Gao [1,2], Jinjun Xu [1,2], Yue Zhu [1], Zhiyan Zhang [3,*] and Quansheng Zeng [1,2,*]

[1] Aviation Mechanical and Electrical Equipment Maintenance College, Changsha Aeronautical Vocational and Technical College, Changsha 410124, China
[2] Aircraft Coating and Structural Repair Laboratory, Hunan Aircraft Maintenance Engineering and Technology Research Center, Changsha 410124, China
[3] Institute of Semiconductors, Chinese Academy of Sciences, Beijing 100083, China
* Correspondence: zyzhang@semi.ac.cn (Z.Z.); zqs_csgx@163.com (Q.Z.)

Abstract: In the field of construction, the surface of architectural aluminum formwork for concrete pouring will remain the concrete adhesion layer of heterogeneous composite structures. In view of the difficulty of removing the concrete adhesion layer, we studied the technology and mechanism of removing the concrete adhesion layer by laser cleaning technology in this paper. We analyzed the composition and distribution characteristics of residual concrete on the surface of architectural aluminum formwork, set up a laser cleaning test system, carried out laser cleaning experiments on the concrete layer on the surface of architectural aluminum formwork under different storage times, and analyzed the mechanism of removing the concrete adhesion layer by laser cleaning. The experimental results showed that the residual time of concrete will affect the quality and efficiency of laser cleaning concrete residue on the surface of architectural aluminum formwork for concrete pouring. For concrete residues with short residual time, lasering can achieve efficient and high-quality cleaning. A nanosecond pulsed laser could strengthen the surface hardness of the aluminum alloy template during cleaning, which is helpful in improving the durability of the aluminum alloy template. The main mechanisms of laser cleaning to remove the concrete adhesion layer on the surface of architectural aluminum formwork is that the bubbles and water bubbles in the loose structure of concrete instantly absorb high-energy laser and make the concrete aggregate continuously air-burst. This paper provides technological and theoretical support for the application of laser cleaning technology to remove residual concrete on the surface of architectural aluminum formwork for concrete pouring in the field of construction.

Keywords: architectural aluminum formwork; concrete adhesion layer; laser cleaning; cleaning mechanism; process experiment

Citation: Gao, K.; Xu, J.; Zhu, Y.; Zhang, Z.; Zeng, Q. Study on the Technology and Mechanism of Cleaning Architectural Aluminum Formwork for Concrete Pouring by High Energy and High Repetition Frequency Pulsed Laser. *Photonics* **2023**, *10*, 242. https://doi.org/10.3390/photonics10030242

Received: 12 January 2023
Revised: 13 February 2023
Accepted: 20 February 2023
Published: 22 February 2023

1. Introduction

In the construction of concrete pouring, aluminum alloy formwork has the advantages of a beautiful appearance, high quality, low average single use cost (more than 300 times of reuse), and greatly shortening the construction period. It has become a representative green infrastructure method with low consumption, environmental protection, economy, and practicality, and is widely used in construction engineering [1,2]. However, due to the strong adsorption and poor ductility of aluminum alloy formwork, it is difficult to remove the residual concrete attached to its surface after construction. Traditional methods [3] such as vibration, pickling soaking, shot blasting, and high-pressure water abrasive washing will damage the surface of the aluminum alloy template, reduce the service life, and also produce a lot of dust and waste liquid, resulting in serious pollution. Therefore, we need an environmentally friendly and convenient cleaning method for residual concrete on the surface of aluminum alloy formwork. It is of great significance to improve the efficiency and quality of construction projects, and to eliminate pollution.

Laser cleaning has been widely used by people since its birth. It has the advantages of being green, having high efficiency, and being easy to control [4]. Since the 1980s, laser cleaning has been successfully applied to the cleaning of dirt on the surface of cultural relics and statues, such as marble, murals, buildings, etc. [5]. In recent years, it has solved the problem that it is difficult to remove tiny contaminated particles with strong adsorption on the mask surface [6] in the field of semiconductor precision manufacturing.

The preferred light source for laser cleaning is a high energy (10–500 mJ) and high repetition frequency (kHz) nanosecond (10–100 ns) pulse laser. This is because the nanosecond narrow pulse width laser acting on the pollutant surface can reduce the heat conduction caused by heat accumulation and avoid excessive temperature rise of the substrate [7–9]. The high repetition frequency pulse acts on the surface of the pollutant, which causes the pollutant to quickly reach the separation and removal threshold, and improves cleaning ability and efficiency [10–12]. High-energy laser pulses can use large focused light spots (400–2000 μm) to remove pollutants, it is beneficial to protect the substrate from damage and reduce the difficulty of cleaning process parameters control.

In recent years, with the rapid development of high-energy and high-repetition frequency nanosecond pulse lasers, the laser cleaning efficiency has been greatly improved, and the application field has been expanding. It has been applied in the fields of mold cleaning, metal welding seam de-gluing, and oxide film removal [13,14], as well as aircraft parts and complete machine paint removal [15,16], warship rust removal [17–19], etc. Traditional cleaning methods are being gradually replaced.

Based on the above background and the development status of laser cleaning technology, laser cleaning technology is applied in the field of construction. In this paper, the laser cleaning technology and mechanism of cleaning the concrete adhesion layer on the surface of aluminum alloy formwork for concrete pouring are studied.

The concrete residue attached to the surface of aluminum alloy formwork is a multi-component mixed structure composed of coarse and fine aggregate (stone, sand) and hydration products of cement. The structural characteristics of the cleaned concrete should be fully considered in the process experiment and mechanism analysis, which are different from the traditional stress stripping and gasification by ablation. Firstly, we analyze the morphology characteristics of the concrete adhesion layer on the surface of aluminum alloy formwork and classify it according to the thickness of the adhesion layer. Secondly, we built a high-energy and high-frequency nanosecond laser cleaning system for aluminum alloy formwork and carried out laser cleaning experiments on concrete adhesion layers with different thicknesses and different residual times. Finally, the cleaning results and the laser cleaning mechanism of concrete with multi-component hybrid structure are analyzed and studied.

2. Characteristic Analysis of Residual Concrete on the Surface of Aluminum Alloy Formwork

After the concrete is poured and the mold removal condition is reached, a certain amount of residual concrete layer will be attached to the surface of the aluminum alloy formwork. It is very important to analyze the microstructure and distribution of the concrete residue layer to study the laser cleaning mechanism of aluminum alloy formwork surface concrete, formulate the process method, and clarify the process parameters.

2.1. Concrete Composition

As shown in Figure 1a, from the macroscopic point of view, concrete is a heterogeneous composite structure. It is mainly composed of coarse and fine aggregates (stone, sand) and hydration products of cement. Aggregate accounts for about 70% of the total volume, cement slurry accounts for about 25%, and there are about 5% pores containing water and gas. The quality characteristics of concrete mainly depend on the properties and relative content of aggregate and cement slurry, and the bonding strength of the interface between aggregate and cement. From the submicroscopic point of view, due to the aggregate gap and "bleeding" caused by particle gradation, there are a large number of porous (30–60 μm) and loose interfacial transition zones between aggregate and mud. As shown in Figure 1b, there are micro-cracks and voids containing a lot of water and air in this area, which is the weak link of concrete strength [20].

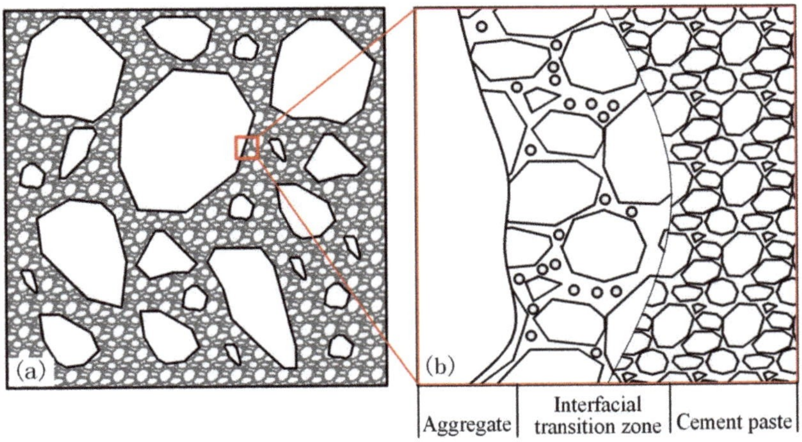

Figure 1. Composition of ordinary concrete structure (**a**) macrostructure; (**b**) submicrostructure.

2.2. Characteristics of Residual Concrete on the Surface of Aluminum Alloy Formwork

The aluminum alloy formwork material commonly used in the construction field for concrete pouring is 6061-T6. The attached concrete is the cement cementitious material and sand aggregate naturally attached during disassembly. The physical properties of the two materials are shown in Table 1. As shown in Figure 2, after the aluminum alloy formwork is removed, the coarse aggregate with a larger particle size is left in the pouring wall or falls off with the removal of the aluminum alloy formwork, while the fine gravel is attached to the surface of the aluminum alloy formwork under the action of cement bonding force.

In this paper, according to the distribution of concrete adhesion layer and mud thickness on the surface of aluminum alloy formwork, the disassembled aluminum alloy formwork surface can be divided into three cases: bare area (residual concrete thickness 10~30 μm), thin slurry area (residual concrete thickness 30~100 μm), and thick slurry area (residual concrete thickness >100 μm). The bare area is mainly composed of isolation glue and dust. There are clear boundaries between each area, as indicated by the yellow arrow in Figure 2.

Table 1. Physical performance parameters of architectural aluminum formwork and its attached concrete.

Materials	Elastic Modulus (MPa)	Coefficient of Linear Expansion (K^{-1})	Poisson's Ratio (υ_a)	Thermal Diffusivity (m^2/s)	Thermal Conductivity (W/(m·K))	Melting Point (°C)
Aluminum formwork 6061-T6	7×10^4	23×10^{-6}	0.3	2.3×10^{-5}	155	580~650
Concrete C15~C80	$2.20\text{~}3.80 \times 10^6$	$4.76\text{~}12.1 \times 10^{-6}$	0.2	1.34×10^{-3}	1.28	1800~2500

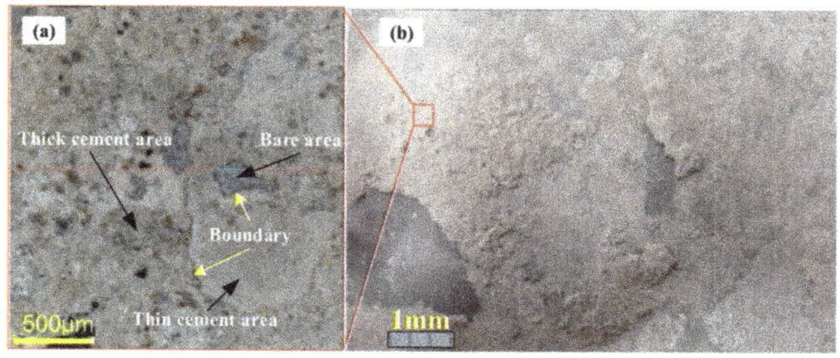

Figure 2. Morphology characteristics of concrete adhered to the surface of aluminum formwork (**a**) microstructure; (**b**) structure.

3. Laser Cleaning Experiment of Residual Concrete on Aluminum Alloy Formwork Surface

3.1. Laser Cleaning Experiment and Testing System

As shown in Figure 3, the laser wavelength of the laser cleaning experimental system is 1 μm, the maximum output power is 600 W, the output frequency is continuously adjustable from 20 kHz to 50 kHz, the pulse width is 100 ns, the maximum single pulse energy is 30 mJ@20 kHz, the diameter of the tail fiber core for laser flexible transmission is 400 μm, and the numerical aperture is 0.2. The two-dimensional motion platform is used to drive the aluminum alloy formwork to translate at a fixed speed, and the positioning accuracy is less than ±20 μm. The paint film thickness gauge (QNIX, 4500, Manufactured by QuaNix in German) is used to detect the thickness of residual concrete on the surface of the aluminum alloy formwork. The measuring range is 0~3000 μm, and the accuracy is ±(3% + 2) μm. The surface cleanliness of aluminum alloy formwork after cleaning is tested by Olympus Digital Microscopy system (OLYMPUS, DSX1000, version: 1.2.2.36). The surface hardness of aluminum alloy formwork after cleaning is tested using an indentation Wechsler hardness tester (HV-1000Z); the measuring range is 0~3000 HV. To accurately detect the change in the thickness of the concrete layer on the surface of the aluminum alloy formwork before and after cleaning, we install a positioning fixture on the moving platform to ensure the positioning accuracy of the repeated installation of the aluminum alloy formwork. The thickness gauge is fixed on the guide slider. It can move along the guide rail perpendicular to the truss to measure the thickness of the concrete layer on the surface of the aluminum alloy formwork.

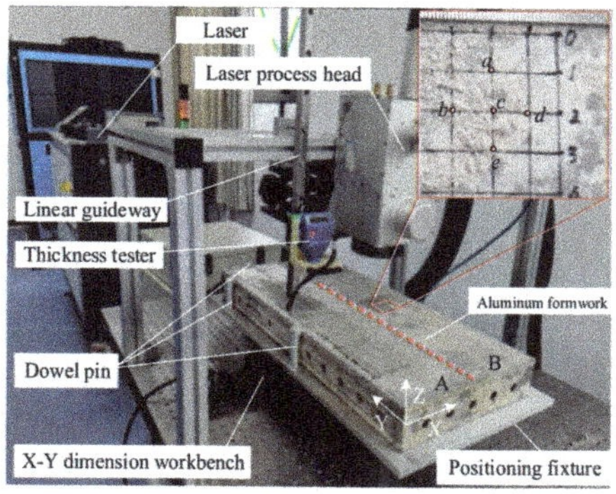

Figure 3. Laser cleaning test and detection system.

3.2. Experimental Sample

In this paper, the laser cleaning experiment is carried out by using the planar aluminum alloy formwork with a concrete layer attached to the surface. The grade of aluminum alloy formwork material is 6061-T6, the size specification is 650 × 300 × 65 mm (L × B × H), and the type of residual concrete on the surface of aluminum alloy formwork is C30.

3.3. Experimental Method

First of all, in order to fully simulate the actual working conditions, the aluminum alloy formwork is divided into two areas, and concrete samples with different residual time are respectively prepared. As shown in Figure 3, the residual time of the concrete residue on the surface of the aluminum alloy formwork in area A is 72 h, and that of concrete residue in area B is 4 weeks.

Secondly, the test area of the residual concrete on the surface of the aluminum alloy formwork is gridded, and the grid size is 10 mm×10 mm. Each grid is numbered and the coordinate direction is marked to form a cleaning interval with three different types of residual concrete layers: thick, thin, and bare. After each positioning, we timely measure and record the height values of four coordinate points a, b, d, and e before and after laser cleaning. The four measuring points are near the center point c, as shown in the aluminum alloy template in Figure 3.

To achieve the highest efficiency of laser cleaning in the practical application of aluminum alloy formwork in concrete pouring, we use the highest power P that the laser can provide. The residual concrete on the surface of the aluminum alloy formwork with a residual time of 72 h or 4 weeks was cleaned by laser. The process standard is to completely remove the concrete layer of a certain state, and the specific process parameters are shown in Table 2. The evaluation indexes are cleaning quality and cleaning time of a single piece, to verify the feasibility of the laser cleaning technology in aluminum alloy formwork cleaning.

Table 2. Technological parameters of laser cleaning experiment.

Laser Power P (W)	Frequency f (kHz)	X-Axis Scanning Width D (mm)	X-Axis Scanning Speed Vx (mm/s)	Y-Axis Speed Vy (mm/s)	Spot Diameter (mm)
600	20	50	2000	10	0.8

The laser energy density and energy value of the high repetition frequency pulse laser acting on the unit area of concrete surface per unit time can be calculated by Equation (1). It is used to formulate the process standard of laser cleaning to remove concrete from the surface of the aluminum alloy mold.

The number of pulses per unit scale of two-dimensional scanning cleaning is:

$$P_n = \frac{n \cdot N}{D}, \tag{1}$$

where N is the number of pulses output by the laser in a single scanning process:

$$N = f \cdot \frac{D}{V_x}, \tag{2}$$

n is the average overlap rate in the Y direction:

$$n = (d/V_y)/(D/V_x), \tag{3}$$

V_x is the one-dimensional scanning speed in the X direction of the galvanometer, V_y is the translation velocity in Y direction of the substrate, D is the scanning range of the galvanometer in the X direction, f is the repetition frequency of pulsed laser, and d is the diameter of the spot.

By substituting the parameters of Table 2 into Equations (1)–(3), we determine that the laser energy per unit area and unit time used in this paper is 114 mJ.

4. Experimental Result

4.1. Effect of Residual Time of Concrete on Cleaning Efficiency

The laser cleaning process parameters in Table 2 are used to clean the surface of aluminum alloy formwork in area A and B, respectively. The two areas A and B have different concrete residual times.

4.1.1. Comparison of the Macroscopic Morphology

The cleaning results are shown in Figure 4a, and the enlarged observation of ①~④ areas in A and B are shown in (b)~(e). Among them, the surface of the aluminum alloy formwork substrate in region A is completely exposed and shows metallic luster. The total cleaning time of area A is about 195 s. The cleaning result of area ① is shown in Figure 4b.

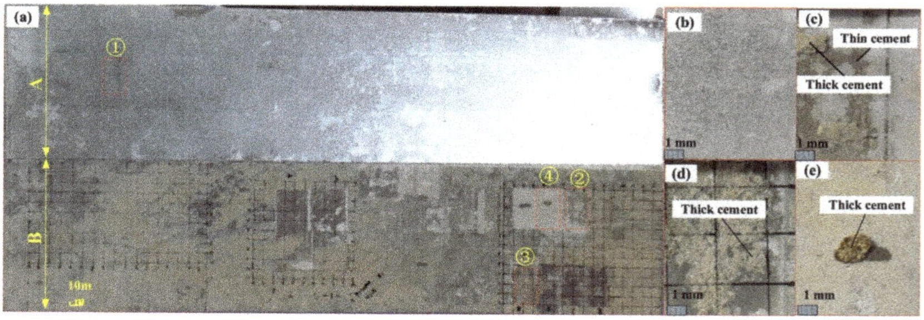

Figure 4. Macroscopic morphology and local magnification of cleaning surface: (**a**) Overall appearance of aluminum formwork; (**b–e**) ①~④ area enlargements.

However, under the same cleaning process parameters, the concrete adhesion layer in area B cannot be completely removed, and there are still more concrete adhesion layers in thin slurry area ② and thick slurry area ③. Area ④ is the surface of thin slurry area ② after repeated cleaning with the same cleaning parameters for five times. Its surface is different from that of area ①. The color of the surface is yellowish and some residual blocks with a black color and a thickness of 300~400 μm are not removed.

4.1.2. Comparison of the Thickness of the Residual Layer on the Cleaning Surface

Table 3 shows the measured values of concrete layer thickness at five specific points before and after laser cleaning on the surface of the aluminum alloy formwork. At position ① in area A, the concrete residue on the surface of both thin and thick slurry areas is almost completely removed, and the thickness of the residual concrete layer is less than 5 μm. The concrete layer at ② and ③ of area B was removed about 20~50 μm after being cleaned by laser for a single time. At position ④ of area B, after repeated cleaning five times, the thickness of the residual block is about 453 μm, and the yellowing layer still has a relatively thick concrete layer, with a thickness of about 8.6~15.6 μm.

Table 3. Concrete height values before and after cleaning in each area (μm).

Area	Measuring Time	a	b	c	d	e
①	before	38.7	27.8	57.1	243	47.9
	after	1.3	2.3	2.1	4.8	1
②	before	46.9	57	47.4	13.9	54.5
	after	13.4	6.4	21.5	2.3	8.7
③	before	780	923	849	408	582
	after	724	892	792	384	342
④	before	54.3	241	470	68.2	34.6
	after	10.2	211	453	15.6	8.6

Through comparative analysis, it can be determined that when the concrete remains on the surface of the aluminum alloy formwork for a short time (<72 h), whether in the thick or the thin slurry areas, using a laser to clean the residual concrete on the surface of aluminum alloy formwork can achieve a better removal effect. However, when the concrete remains on the surface of the aluminum alloy formwork for a long time (4 weeks), laser cleaning of residual concrete on the surface of the aluminum alloy formwork cannot achieve a good removal effect in both thick and thin slurry areas. It specifically shows a low cleaning efficiency and more surface residue after cleaning. From the perspective of engineering application, the residual time of 72 h selected in this paper can meet the time requirements of laser cleaning at the construction site. The selected time also meets the time requirements for transferring it back to the aluminum formwork factory for laser cleaning. Under the experimental conditions of this paper, it is estimated that the cleaning time for one aluminum alloy formwork is 6.5 min, which has high feasibility in engineering application. This cleaning efficiency has high feasibility in engineering application.

4.2. Comparison of Micro-Morphology

4.2.1. The Surface Morphology of Area A after Being Completely Cleaned

Figure 5a shows the typical surface micro-morphology at ① in area A after complete cleaning. The surface concrete is almost completely removed, and the metallic luster of the aluminum alloy formwork substrate is completely exposed. As shown in Figure 5b, the microscopic surface of the aluminum alloy formwork is covered with micro-pits of different sizes and depths. The maximum depth of the pits is about 40 μm, which is the surface damage caused by concrete gravel extrusion and early high-pressure water abrasive washing during the application of the aluminum alloy formwork. Figure 5c shows the exfoliation collected after laser cleaning. The exfoliation is a micro flake structure containing fine aggregate. After absorbing a high-energy laser, the temperature of concrete sharply rises, which forms a higher temperature gradient with the aluminum alloy substrate, and forms thermal stress. However, the isolation agent between concrete and aluminum alloy has no bonding effect, and the connection strength is much less than the thermal stress, which leads to a regular peeling off of the exfoliation.

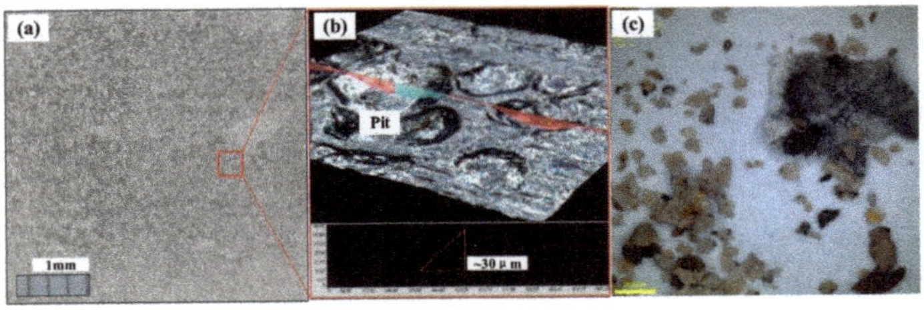

Figure 5. Typical surface micro-morphology after cleaning at ① in area A: (**a**) 2D view; (**b**) 3D view and measured value; (**c**) the exfoliation collected after laser cleaning.

4.2.2. The Surface Topography of Area B That Cannot Be Cleaned

Figure 6 shows the surface morphology at ②~④ in area B that have not been completely cleaned. By observing Figure 6a,b, it can be found that before the surface residue of the thin slurry area ② is cleaned, some of the metal substrate is exposed (indicated by the red dot arrow), and the attachment layer in some areas is not completely peeled off. Due to the isolation agent, the residue showed warped scales. The micro-morphology after removal of the residue shows that the thin residual layer with cement as the main body is still attached, and the exposed metal area and the residual area are intertwined and distributed. According to Figure 6c,d, it can be found that no obvious cleaning exfoliation can be found on the surface of the concrete layer in the thick slurry area ③, and there are a large number of pits (indicated by the yellow arrows) and semi-exposed tiny aggregate (indicated by the red arrows) on the surface. The depth of the deeper pit is about 110 μm.

However, in the range of visual distance, there is one fresh pit, and the surface color of the other pits is the same as that of the protruding surface. According to Figure 6e,f, it can be found that the height of the residual block at ④ in area B after repeated cleaning is about 360~480 μm, and its surface color is darker. There is one fresh pit (indicated by the yellow arrow) within the visual range, while the other pits have a similar surface color to the protruding surface.

Figure 6. Typical micro-morphology at ②~④ in area B: (**a**,**b**) morphology at ②; (**c**,**d**) morphology at ③; (**e**,**f**) morphology at ④.

By comparing the above results, it can be concluded that the laser can effectively clean the concrete on the surface of the aluminum alloy formwork for architectural concrete pouring that has been dismantled in a short period of time. However, when the concrete remains on the surface of the aluminum alloy formwork for a long time, the removal efficiency and quality of laser cleaning are low.

4.3. Surface Hardness Test of Aluminum Alloy Formwork after Laser Cleaning

As shown in Figure 7a, the surfaces adjacent to areas A and B are selected to test the hardness of the cleaned and uncleaned but completely exposed aluminum alloy formwork substrate. The measured values are shown in Figure 7b. The results show that under the action of instantaneous thermal shock of the pulsed laser, the surface of the aluminum alloy template is hardened to a certain extent, and the surface hardness is slightly improved. It is beneficial to improve the wear resistance and service life of the aluminum alloy formwork.

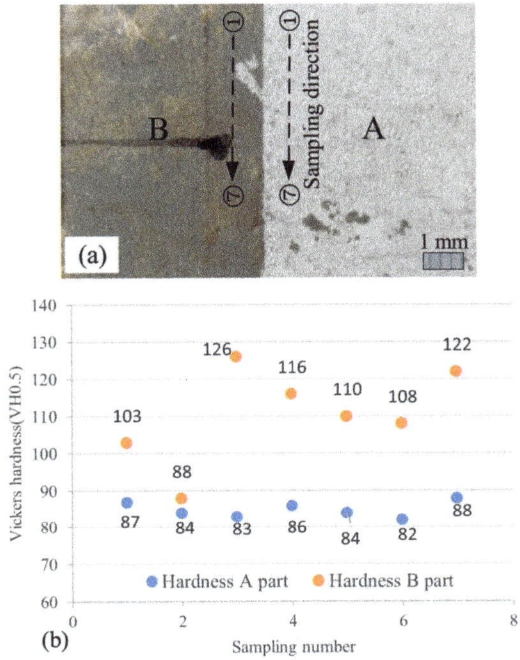

Figure 7. (**a**) Vickers hardness test sample; (**b**) Hardness test result.

5. Mechanism Discussion

5.1. Laser Cleaning Mechanism of Thick Concrete Slurry Layer

As shown in Figure 8, when the laser hits the surface of the concrete layer, the tiny aggregates and cement compounds in the concrete absorb the laser energy at the same time. The thermal conductivity of concrete is poor, and the local temperature of the irradiated area rapidly rises. However, it is not enough to make high temperature resistant aggregate (main component SiO_2) and cement slurry ablation and gasification. Therefore, concrete cannot be removed by ablation and gasification as easily as paint and other organic matter. However, there is an interface transition zone between concrete aggregate and cement slurry, and its structure is loose and there are many air bubbles and water bubbles. Their volume violently expands under high temperature, which leads to explosions and impact peeling. The tensile strength of concrete is only 1/10 of the compressive strength. Under the above impact, the tiny aggregate and mud all break away and fly out, forming fine dust to achieve the purpose of cleaning and removal.

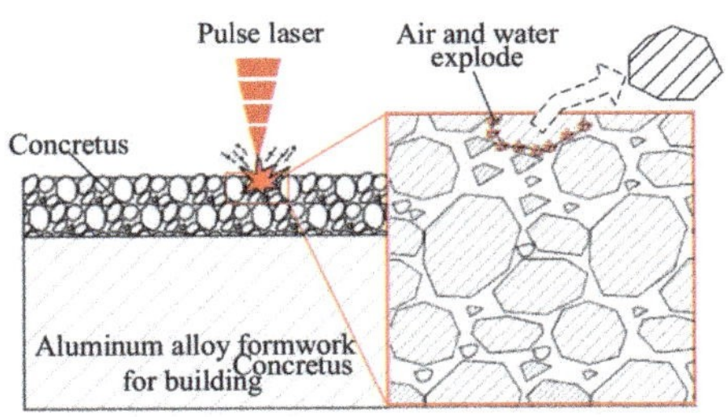

Figure 8. Laser cleaning mechanism of thick concrete slurry layer.

After the concrete layer is placed for a long time, the water content of the loose tissue of the transition layer decreases or evaporates. The insufficient energy of steam explosion caused by air bubbles heating can only cause a small amount of aggregate with a weak binding force to explode, crush, and peel, as shown in Figure 6c–f.

Therefore, the air bubbles and water bubbles in the transition interface layer between the concrete aggregate and the cement slurry instantly absorb high-energy laser gasification explosion and form an impact pressure wave. This is the main mechanism of laser cleaning to remove the thicker concrete slurry layer.

5.2. Laser Cleaning Mechanism of Thin Concrete Slurry Layer

As shown in Figure 9, the residual thin paste layer of concrete is mainly composed of cement paste and smaller aggregates dispersed in it. After laser irradiation, the main cleaning mechanism of aggregate and cement slurry is that the air bubbles and water bubbles in the interface transition zone are gasified at a high temperature, and finally the smaller aggregate is removed by blasting impact. Because the cement layer is thin, the impact of blasting will cause cracks in the thin cement layer and transfer along the surface of the aluminum alloy substrate. It can force the cracked cement layer to leave the surface of the aluminum alloy substrate with the isolator, forming a flake cleaning material collected from Figure 5c.

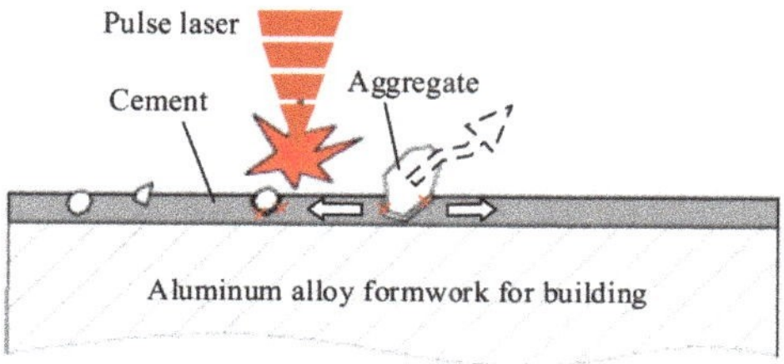

Figure 9. Laser cleaning mechanism of thin concrete slurry layer.

However, the thin slurry layer, which has been placed for a long time, has less water content and insufficient impact force of an air steam explosion, so it is difficult to produce enough cracks in the thin slurry layer. At the same time, the cement layer is resistant to high temperatures and cannot be directly removed by gasification after laser action. The poor thermal conductivity of cement leads to the deterioration of laser thermal vibration, which can only form the incomplete separated scale residue shown in Figure 6a.

In summary, the concrete layer mainly consists of aggregate and cement, and their interface transition area has a loose structure and low strength. The shock wave formed by the gasification explosion of air bubbles and water bubbles in this area after absorbing high-energy laser is the main mechanism of cleaning architectural aluminum alloy formwork by laser.

Figure 10 is a mechanism diagram of the laser cleaning and removal of multi-component structure materials represented by concrete structures. The multi-component structure substances contain materials with low vaporization temperature or high thermal expansion coefficient, which rapidly expand and deform or form a blast after absorbing a high-energy laser. Therefore, based on this effect, we can add water and other substances that are easy to vaporize and form explosions when cleaning some materials, so as to improve the cleaning efficiency and ability.

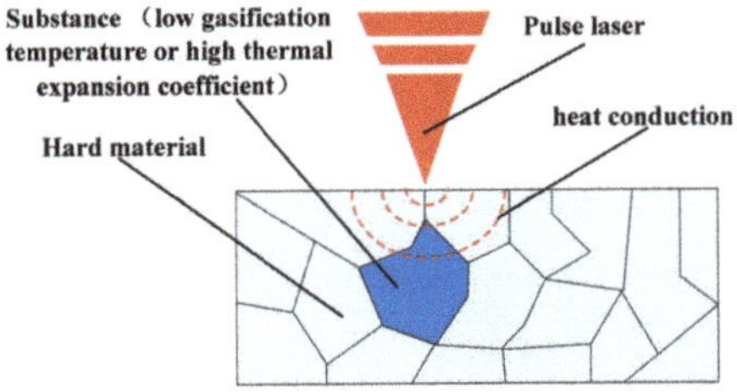

Figure 10. Mechanism diagram of laser cleaning materials with multi-component structures containing substances with low vaporization temperatures or high thermal expansion coefficients inside.

6. Conclusions

In this paper, based on the laser cleaning device built by a high-energy and high-frequency nanosecond pulse laser, we carried out laser cleaning experiments on the residual concrete layer on the surface of the aluminum alloy formwork with different thicknesses and disassembly times. Finally, the following conclusions were obtained:

(1) The residual time of concrete residue on the surface of aluminum alloy formwork for concrete pouring will affect the quality of laser cleaning. For concrete with a short residual time, lasering can realize high-efficiency and high-quality cleaning. Using the experimental system and process parameters in this paper, the cleaning rate was 6.5 min per formwork, which has high engineering application feasibility. However, for concrete with a long residual time, the efficiency and quality of laser cleaning were poor.

(2) The nanosecond pulse laser will not cause secondary damage to the surface of the aluminum alloy formwork. It can strengthen the hardness of the surface of the aluminum alloy formwork. After laser cleaning, the surface hardness of the aluminum alloy formwork is slightly improved, which helps to improve the durability of the aluminum alloy formwork.

(3) The air bubbles and water bubbles in the loose concrete structure instantaneously absorb the high-energy laser, causing the concrete aggregate to continuously explode. This is the main mechanism for laser cleaning to remove the concrete adhesion layer on the surface of aluminum alloy formwork.

Author Contributions: Conceptualization, Q.Z. and Z.Z.; methodology, K.G.; investigation, J.X.; writing—original draft preparation, Y.Z. All authors have read and agreed to the published version of the manuscript.

Funding: This research was funded by "Hunan High-Tech Industry Science and Technology Innovation Leading Plan, grant number 2020SK2027" and The Laser system was funded by "National Key R&D Plan, grant number 2022YFB4601501".

Data Availability Statement: The data that support the findings of this study are available from the corresponding author upon reasonable request.

Conflicts of Interest: The authors declare no conflict of interest.

References

1. Wang, Y.H.; Li, Q.Z. Application of Architectural aluminum formwork in Some Super Tall Building. *Constr. Technol.* **2011**, *40*, 35–37+75.
2. Zhang, A.Q. Research and Application of Aluminum Formwork Technology in Super Tall Building. Master's Thesis, Changchun Institute of Technology, Changchun, China, 2016.
3. Li, Y. A pass-through cleaning system of aluminum alloy template. *Clean. World* **2019**, *35*, 10–11.
4. Song, F.; Liu, S.J. Principle and application research on laser cleaning. *Clean. World* **2005**, *21*, 1–6.
5. Pleasants, S.; Kane, D.M. Laser cleaning of aluminum particles on glass and silica substrates: Experiment and quasistatic model. *Appl. Phys.* **2003**, *93*, 8862–8866. [CrossRef]
6. Cooper, M.; Larson, J. The use of laser cleaning to preserve patina on marble sculpture. *Conservator* **1996**, *20*, 28–35. [CrossRef]
7. Qiu, T.W.; Yi, J.L. Characteristics of Nanosecond Pulse Laser Cleaning Paint Coatings on 2024 Aluminum Alloy Surface. *Laser Optoelectron. Prog.* **2021**, *58*, 0514001.

8. Guo, L.; Li, Y. Numerical and experimental analysis for morphology evolution of 6061 aluminum alloy during nanosecond pulsed laser cleaning. *Surf. Coat. Technol.* **2022**, *432*, 128056. [CrossRef]
9. Zhang, D.; Xu, J. Removal mechanism of blue paint on aluminum alloy substrate during surface cleaning using nanosecond pulsed laser. *Opt. Laser Technol.* **2022**, *149*, 107882. [CrossRef]
10. Qi, X.S.; Ren, Z.G. Study on effect of laser derusting technology on properties of high speed train collector ring. *Laser Technol.* **2019**, *43*, 168–173.
11. Shi, S.D. Researches on Theoretical Model, Numerical Simulation and Application of Painting Removal by Laser Cleaning. Doctor's Thesis, Nankai University, Tianjin, China, 2012.
12. Zhang, Z.Y.; Wang, Y.B. Removal of Low Thermal Conductivity Paint with High Repetition Rate Pulse Laser. *Chin. J. Lasers* **2019**, *46*, 148–156.
13. Lei, Z.; Tian, Z. Laser Cleaning Technology in Industrial Fields. *Laser Optoelectron. Prog.* **2018**, *55*, 60–72.
14. Zhang, G.; Hua, X. Investigation on mechanism of oxide removal and plasma behavior during laser cleaning on aluminum alloy. *Appl. Surf. Sci.* **2020**, *506*, 144666. [CrossRef]
15. Zhu, G.D.; Wang, S.R. Corrosion and wear performance of aircraft skin after laser cleaning. *Opt. Laser Technol.* **2020**, *132*, 106475. [CrossRef]
16. Ünaldi, S.; Papadopoulos, K. Towards selective laser paint stripping using shock waves produced by laser-plasma interaction for aeronautical applications on AA 2024 based substrates. *Opt. Laser Technol.* **2021**, *141*, 107095. [CrossRef]
17. Gao, K.; Zeng, Q.S. Adhesion of Residual Primer Paint After Laser Cleaning Aircraft Aluminum Alloy Skin. *Laser Optoelectron. Prog.* **2021**, *58*, 296–302.
18. Zheng, G. Study on Laser Paint Stripping of Aircraft Surface. Master's Thesis, Institute of Electrics, CAS, Beijing, China, 2005.
19. Li, J.M. Development, Trend and Application of High Average Power Diode Pumped Lasers. *Laser Optoelectron. Prog.* **2008**, *45*, 16–29.
20. Zhang, F.Q. *Concrete Material*; Chemical Industry Press: Beijing, China, 2015.

MDPI
St. Alban-Anlage 66
4052 Basel
Switzerland
www.mdpi.com

Photonics Editorial Office
E-mail: photonics@mdpi.com
www.mdpi.com/journal/photonics

www.ingramcontent.com/pod-product-compliance
Lightning Source LLC
Chambersburg PA
CBHW061630210926
51850CB00009B/2503
9783036587134